Writing Papers in the Biological Sciences

Writing Papers in the Biological Sciences

Third Edition

Victoria E. McMillan
Colgate University

BEDFORD / ST. MARTIN'S   Boston ◆ New York

For Bedford/St. Martin's

Developmental Editor: Jennifer Rush
Production Editor: Stasia Zomkowski
Production Supervisor: Sarah Zwiebach
Marketing Manager: Brian Wheel
Editorial Assistant: Caroline Thompson
Copyeditor: Alison Greene
Text Design: George McLean
Cover Design: Hannus Design Associates
Cover Art: Cover Photograph: Broadtail Hummingbird at Honeysuckle.
© W. Shattil/B. Rozinski. Courtesy of Natural Selection Stock.
Composition: Pine Tree Composition, Inc.
Printing and Binding: R.R. Donnelley/Crawfordsville

President: Charles H. Christensen
Editorial Director: Joan E. Feinberg
Editor in Chief: Karen S. Henry
Director of Marketing: Karen Melton
Director of Editing, Design, and Production: Marcia Cohen
Managing Editor: Elizabeth M. Schaaf

Library of Congress Catalog Card Number: 00-106456

For information, write: Bedford/St. Martin's, 75 Arlington Street, Boston, MA 02116 (617–399–4000)

ISBN: 0–312–25857–7

Preface

As a biologist who also teaches composition, I see many biology students who, although they have strong backgrounds in biology, have trouble understanding the writing tasks and challenges unique to their discipline. Biologists need to think creatively about science, and they need to write about science clearly, accurately, and concisely. In fact, these two processes, thinking and writing, are as interdependent in science as they are in any other field. Unfortunately it is difficult to fit any extensive discussion of scientific writing into a regular semester-long biology course. I hope this book can substitute for such discussion, serving as a supplemental text for courses requiring laboratory reports or longer written assignments. Although intended primarily for undergraduates who need an introduction to the aims and format of biological writing, it will also be useful for more experienced students, including master's and doctoral candidates who are preparing dissertations or manuscripts for publication. Finally, I hope this book will also find its place in the small but growing number of courses on scientific writing at the undergraduate or graduate level.

Like its two predecessors, this edition is organized as a self-help manual, offering straightforward solutions to common problems and numerous examples of both faulty and effective writing. All of the book's most important features—rules highlighted with bullets, numerous examples, clear and concise explanations, and a detailed index—have been retained, including the spiral binding for easier use. The third edition includes expanded discussions of many topics as well as much new material. The Introduction discusses the role of both formal and informal writing in biology and gives an overview of the diversity of writing tasks faced by professional biologists. Chapter 1, a detailed guide to locating and using biological literature, has been thoroughly revised to address the ever-changing parameters of electronic research. Coverage includes guidelines for conducting online searches, updated descriptions of a wide variety of scientific databases, and tips on evaluating Internet sources. Sections on how to read scientific papers, how to take good notes, and how to avoid plagiarism—always of concern to students—have been retained.

Chapters 2 and 3 provide an introduction to the tasks of gathering, analyzing, and presenting data. New to Chapter 2 is a discussion of how to use spreadsheets in recording data. Chapter 3 now includes updated examples of figures and tables. Both chapters still include advice on experimental design and the use of statistics both in writing and in research.

Chapters 4 and 5 discuss the two types of written assignments most commonly faced by biology students. Chapter 4 focuses on the *research paper*, the primary form in which professional biologists communicate original findings. Students are expected to follow this model when they prepare research projects or laboratory and field reports. Chapter 5 discusses the *review paper*, which resembles the library-oriented "term paper" assignments of advanced courses. Both chapters include many examples, as well as an annotated sample paper—a new one in Chapter 4.

Students often get confused about how to use literature citations and how to prepare the Literature Cited section of their papers. Chapter 6 discusses the CBE (Council of Biology Editors) format for citing and listing references, along with examples of common variations of the CBE format. Coverage also includes guidelines for citing electronic sources. CBE format is elucidated by each of the sample papers in Chapters 4 and 5.

Many students are uncertain about how to begin a paper or how to revise it effectively. Chapter 7 discusses the process of drafting and revising, and presents the basics of each stage of the writing process—from brainstorming to composing the first draft, to final editing at the paragraph and sentence levels. To help students prepare their final drafts, Chapter 8 discusses manuscript format, provides proofreading strategies, explains how to write an Acknowledgments section, and lists symbols and abbreviations commonly used in biology.

Students often underestimate the role that writing can play in learning difficult material and in studying for examinations. Chapter 9 offers strategies for taking notes, getting the most out of textbooks, studying for short-answer questions, and preparing for laboratory exams. New to this chapter is a section on how to make effective laboratory drawings, with examples.

Finally, this third edition includes updated coverage of other forms of biological writing often faced by biology students. Chapter 10 presents strategies for preparing research proposals, letters of application, résumés, curricula vitae, and both oral and poster presentations.

Throughout the book I have tried to address the special demands and challenges of academic assignments in the context of how and why professional biologists write. I hope this handbook will convince students that scientific writing, including their own, need not be tedious and cumbersome, but can be clear, crisp, incisive, and engaging.

Acknowledgments. I owe many people thanks for their help with the third edition of this book. Deborah Huerta, Science Librarian at Colgate University and consultant for Chapter 1, provided invaluable advice and information, without which the updated material on searching the literature and evaluating Internet sources would not have been possible. Robert Arnold, of Colgate's Biology Department, served as consultant for Chapters 2 and 3 and spent many hours helping to update the tables and figures and to create the new spreadsheet in Chapter 2. I am also grateful to Colgate students Tara Gupta, Brian Martin, Andrew Rozbruch, and Kristin VanderPloeg for giving me permission to adapt their writing for use in this book. I thank William Oostenink for use of his photomicrograph of smilax in Chapter 9 and Judy Fischer for advice about résumés and curricula vitae.

This edition profited from the thoughtful comments of the following reviewers: William Eisenger, Santa Clara University; Marcia Harrison, Marshall University; Terrence McGlynn, Gettysburg College; Laura Mudd, Barry University; Janet L. Pease, Syracuse University; Marianne Robertson, Milliken University; Carl Schaefer, University of Connecticut; Rob Snetsinger, Queen's University; Nina Zanetti, Siena College. In addition, I thank William S. Brown of Skidmore College and the State University of New York at Albany, for numerous suggestions for improvement.

At Bedford Books, I owe many thanks to Publisher Chuck Christensen and Editoral Director Joan Feinberg for their continuing commitment to this book. I also remain grateful to Steve Scipione for guiding me expertly through the first edition, and to Laura Arcari, who saw many possibilities for innovations and additions as I prepared the second edition. The third edition would not have been possible without the editorial guidance of Karen Henry, Editor in Chief, Boston, and Jennifer Rush, Development Editor, and I am deeply grateful for their encouragement, patience, and many excellent suggestions. I am also indebted to Stasia Zomkowski, Production Editor, for invaluable help during the final stages of turning manuscript into print. I thank Kristen Harvey for preparing the tear sheets used for revision and Caroline Thompson for help in revising the list of Additional Readings.

Over the years, many past and present colleagues in Colgate's Interdisciplinary Writing Department—Margaret Darby, Vincent DiGirolamo, David Hess, Rebecca Howard, Sandra Jamieson, Mary Joy, Bruce Pegg, Mary Lynn Rampolla, and Elayne Zorn — have contributed to my growth as a writing instructor and in this way have enriched all editions of this book. Finally, I owe special thanks to Robert Arnold for his continued help and encouragement along the way.

Contents

Writing Papers
in the
Biological Sciences

How and Why Biologists Write: An Introduction to Biological Literature

There are several common misconceptions about the role of writing in science. One of these is that scientists don't actually do much writing; consequently, biology students, unlike students in the humanities, for example, need not worry about their writing abilities. Another misconception is that scientific writing is, at best, dry and cumbersome; at worst, it is tedious, pedantic, and impenetrable. Scientists are rarely credited with the ability or the motivation to write with creativity and flair, and the diversity of writing tasks faced by most biologists is often underestimated.

This book, I hope, will help dispel such misconceptions. Science is a collective enterprise: its growth depends on the work and insight of many individuals and on the free exchange of data and ideas. Biologists *must* be effective writers because no experiment, however brilliant, can contribute to the existing fund of scientific knowledge unless it has been described to others working in the same field. For this reason, the primary aim of scientific writing is to communicate — as clearly, accurately, and succinctly as possible. This goal is not easy to achieve, and examples of stilted, jargon-ridden papers do appear in the scientific journals. However, there are also many examples of good, even brilliant, scientific writing whose authors have managed to translate complex data and theory into clear, straightforward, illuminating prose. Such works have their own kind of grace and elegance.

Most biologists routinely engage in a variety of writing tasks. Their professional work is typically published in scientific journals and is aimed at a specialized group of readers: generally other scientists with some background in the particular subject area. Biologists write two major types of papers for such journals: the research paper, or "scientific paper," and the review paper.

The *research paper* reports original findings from a field or laboratory investigation that may have extended over many weeks, months, or even years. It is generally divided into six distinctive sections: Abstract or Summary, Introduction, Materials and Methods, Results, Discussion, and Literature Cited (sometimes called References). Although editorial requirements vary from journal to journal and authors must conform to those specifications, most scientific papers follow the same basic organization. The format has been dictated by a long history of printing and publishing traditions that reflect the logic and elegance of the scientific method.

The format of a research paper reflects a scientist's obligation to make his or her assumptions clear, the methods repeatable, and the interpretations clearly separate from the data or results. Although it may contain lengthy descriptive data or complex statistics, a research paper is at heart a well-structured *argument* founded on supporting evidence. It serves as a vehicle for presenting one's own findings and conclusions, and for arguing for or against competing hypotheses. If independent investigators use similar methods but come up with different results, then the conclusions of the original study may need reevaluation or modification. In any case, our understanding of the subject will eventually be expanded.

In more practical terms, research papers serve as a principal means by which a particular scientist's contributions are evaluated by his or her peers. As professional biologists know all too well, it is not easy to turn raw data into a published research paper in a journal of one's choice. Most papers have gone through multiple drafts and often extensive revisions before appearing in their published form. Typically, the author solicits comments from his or her colleagues before submitting the paper for publication. In addition, the journal editor sends the manuscript to several reviewers — other scientists in the author's field — who read it critically and recommend that it be either accepted or rejected. Even if accepted for publication, a manuscript usually needs some changes, often major ones, before finally finding its way into print. The entire editorial process takes many months to complete.

Several other, related forms of writing can also convey a biologist's data or ideas. In addition to full-length research papers, many journals also publish shorter reports of original findings. They often appear in a separate section titled Research Notes, Communications, Short Communications, or the like. The organization of such works is similar to that of research papers, although certain sections may be streamlined or abbreviated to conform to shorter manuscript specifications.

A *conference presentation* is a paper presented either orally or in poster format at a formal meeting of scientists interested in a particular field of study. Conference papers present unpublished research, often work in progress; biologists depend on them to learn about the current activities of their colleagues. The abstracts of presentations are usually distributed in printed form to all people registered for the conference; the proceedings of major symposia are listed in major indexes and abstracts, and can be obtained through the facilities of research libraries. Typically, authors of conference talks and posters publish their work as journal papers as soon as possible.

A *research proposal,* or grant proposal, is a carefully organized plan for future research. Biologists submit such proposals to particular committees or agencies — for example, the National Science Foundation — in the hope of obtaining funding for their work. Biology students may also write research proposals in anticipation of independent coursework or Ph.D. research. Because they suggest, rather than present, original studies, research proposals may have lengthy introductions that review existing literature on the topic and provide a rationale for the new work being proposed. Space is also devoted to methodology and projected costs for equipment, research assistants, travel, or other expenses. A results section is lacking, of course, although the introduction may showcase the author's previous findings, particularly if they were obtained with funds from the same agency.

In contrast to research papers, conference presentations, and proposals, a *review paper* is a journal article that synthesizes work by many independent researchers on a particular subject or scientific problem. By bringing together the most pertinent findings of a large number of studies, a review paper serves as a valuable summary of research. Although it does not present the writer's new discoveries, it does reflect his or her painstaking review of the literature in a defined field. Moreover, a good review not only summarizes information but also provides interpretive analysis and sometimes a historical perspective. Reviews vary in aims, scope, length, and format, but they all include a relatively lengthy reference section. Journal editors sometimes invite prominent experts to write reviews of their particular fields, since the ability to give an authoritative overview of a subject usually develops with experience. Whether solicited or unsolicited, review papers still must conform to journal specifications, and their authors receive feedback from editors and reviewers before final publication.

In addition to writing for specialized scientific audiences, many biologists also write extensively for readers with little background in science. For example, magazines such as *Natural History,* published by the American Museum of Natural History, and *Smithsonian,* published by the Smithsonian Institution, feature a wide variety of essays and articles written by experts for nonspecialists. Many scientists, in fact, take great pleasure in transforming their professional work into writings that reach out to the general reader. In addition to magazine submissions, some biologists write books on natural history, health, the environment, or the practice of science; others may compose poetry or write novels. Still others combine science with their involvement in teaching through the writing of textbooks (like this one), or articles in journals such as *Journal of College Science Teaching* or *The American Biology Teacher*.

Biologists who teach at colleges and universities routinely produce a variety of other written products: lectures and course syllabi, laboratory and paper assignments, examinations, letters of recommendation for students and colleagues, memos and departmental reports, and, at times, their own résumés and letters of application for new academic positions. Finally, remember that most *formal* writing is preceded and molded by *informal* writing in the form of notes, outlines, journal entries, laboratory or field data, and communication with colleagues through letters, professional newsletters, and electronic mail and Internet discussion groups. In short, good writing skills are as indispensable to biologists as they are to scholars in other academic disciplines.

CHAPTER 1

Locating and Using Biological Literature

Review paper assignments, research projects and proposals, and many laboratory reports will require you to immerse yourself in biological literature. All scientific endeavors, including your own, acquire their fullest meaning and significance when viewed in the broader context of what other scientists, past and present, have thought and done. Therefore, to be a successful biology student and, with time, an effective biologist, you will need to develop competent information-literacy skills. In other words, you will need to know how to locate the scholarly work of others, how to critically evaluate these sources, and how to incorporate this information effectively into your own writing. The guidelines below will help you get started.

SEARCHING THE LITERATURE

■ Understand the difference between primary and secondary sources.

In the sciences, *primary sources* are reports of original findings and ideas. These generally take the form of peer-reviewed research papers in scholarly journals and are directed toward a scientific audience. Thousands of scientific journals are published around the world. Some, such as *Science* and *Nature,* report findings from diverse branches of science; others, such as *Neurology, Phycologia,* and *Journal of Bacteriology,* are focused more narrowly on particular subdisciplines. Research papers in journals usually deal with relatively specialized topics — for example, "Osmoregulation in the Freshwater Sponge, *Spongilla lacustris*"; "Pelagic Carbon Metabolism in a

Eutrophic Lake During a Clear-water Phase"; or "Facilitated Nuclear Transport of Calmodulin in Tissue Culture." Sometimes a research paper reappears later in book form, as part of an edited collection of noteworthy articles on a particular subject. Although many journal articles are written in English, and may be accompanied by an abstract in another language, such as French, German, or Japanese, others may be written in a language other than English but accompanied by an abstract in English.

Aside from research papers, other primary sources are conference papers, dissertations, and technical reports of government agencies or private organizations. Your clue to recognizing primary literature is usually the presence of a Materials and Methods section, or at least some mention of methodology, as well as a Results section, typically with accompanying tables or figures. These components indicate that the author is presenting new data and ideas.

Secondary sources are more general works that are *based* on primary sources. Many are written for readers with little knowledge of the sciences. These include books on natural history, the environment, and other scientific subjects, as well as articles in magazines such as *Discover, Natural History, Smithsonian, American Scientist, Audubon, Scientific American,* and *National Wildlife*. Although many secondary sources are written for both scientists and nonscientists, others attract a more specialized audience; examples are scholarly books (monographs) and review papers that summarize and interpret the primary literature in a particular subject area. Finally, note that secondary sources, like primary sources, may appear in both print and digital formats.

■ Start by consulting general references.

Before plunging into the specialized primary literature on your topic, make sure you have a solid background and some sense of your aims and scope. For example, if you are writing a paper on regeneration in planarians for an invertebrate zoology course, you might begin by reading all relevant sections of your own course text. Next, you might search for other general books on the biology of flatworms, or on regeneration in invertebrates. Such references will give you some grounding in the topic and may contain extensive specialized bibliographies. Scientific encyclopedias also make good introductory reading and usually contain additional references; these and other books can be located through your library's catalog. If a subject search on your topic reveals few items, ask the reference librarian for help.

Popular science magazines provide accessible introductions to difficult biological material. Their articles give up-to-date overviews of particular subjects and may mention names of important researchers. Their nonspecialist language and approach will prepare you for the more specialized vocabulary of scientific journals — which ultimately should be your major reading material. Many authors of nontechnical articles have also published scholarly papers on the same topic and later you may wish to locate those papers, too.

Review papers provide excellent beginning reading. Although written for scientific audiences, reviews may not require as much background in the specific subject area as do the more specialized research papers they discuss. They give you some historical perspective, summarize the contributions of influential researchers, and provide a sense of the important scientific issues being addressed. Their Literature Cited section will be a rich source of primary literature to consult once your own focus has narrowed. Published reviews are also appropriate models to consult if you are preparing your own review or term paper (see Chapter 5).

Although some journals, such as *Genetics,* specialize only in research papers, many other publications are devoted exclusively to reviews. The latter include *Annual Review of Ecology and Systematics, Advances in Ecological Research, Recent Progress in Hormone Research, Annual Review of Genetics,* and numerous others. Still others, such as *Biological Bulletin,* publish mostly primary literature but also feature occasional review papers.

Finally, do not overlook the usefulness of scholarly monographs, which provide a more extensive overview of the literature than is possible in a review paper, and serve as excellent introductions to more specialized scientific topics. Your reference librarian can assist you in locating such references.

■ Learn to use scientific abstracts and indexes.

Most libraries have a variety of general and specialized indexes and abstracting services. These provide access to the primary and secondary literature in journals, as well as to monographs, dissertations, conference proceedings, documents, and other reports. Both indexes and abstracts contain lists of sources; abstracts also provide a short summary (or *abstract*) of most or all sources they include. Originally, indexes were in print, but now most are available as electronic databases. Each index or abstract has its own method of organization; for print versions, you can easily understand the format if you spend a little time reading the instructions section for users. Similarly, you will need to learn the different protocols for electronic databases by accessing the detailed Help screens. In either case, the most efficient strategy is to consult your reference librarian.

Following is a selection of indexes and abstracts most useful to biologists:

General Science Abstracts is an excellent place to begin your literature search, as it provides access to a wide variety of leading science journals and magazines, including those covering biology, environmental issues, and health.

A somewhat more specialized database is *Biological and Agricultural Abstracts.* This database is useful for topics in environmental and conservation sciences, agriculture, forestry, veterinary medicine, and applied and field biology. It lists articles in selected general journals (for example, *Heredity, The American Naturalist,* and *Ecological Monographs*) as well as in more specialized periodicals (*Crop Science, Tropical Agriculture, Ecological Entomology,* and *Animal Science*). Beginners find it especially useful because it includes many periodicals devoted solely to reviews.

AGRICOLA, the National Agricultural Library's database, covers a greater range of similar subjects and also provides access to more specialized books, journal articles, and documents. Subjects include agriculture, natural resources, aquaculture, plant and animal sciences, forestry, entomology, nutrition, and environmental issues, among others.

CAB International, another comprehensive database for agricultural topics, is available in various digital formats. Topics include all fields of agriculture, nutrition, animal health, forestry, and management of natural resources, as well as related fields of inquiry.

Biological Abstracts (BIOSIS Previews) is the largest life sciences journal literature database. Its manufacturer, BIOSIS, also markets a companion database, *Biological Abstracts/RRM,* that indexes reports, reviews, and meetings. In addition, BIOSIS offers the venerable database *Zoological Record,* originally published by the Zoological Society of London and the British Museum. These three databases are available in print, on CD-ROM, online, and on the Web through a number of vendors. BIOSIS has also signed an agreement with the Institute for Scientific Information (ISI) to make its online database, BIOSIS Previews, interface with *Web of Science* (*Science Citation Index*). This arrangement will allow links to those full-text documents that are available through *Web of Science* provided that your library subscribes to these services. All these databases provide full information on references, including abstracts.

Cambridge Scientific Abstracts also offers a number of fine biology databases. In its *Biological Sciences* database are numerous subfiles for such fields as biology, genetics, ecology, pharmacology, medicine, and bioengineering. *Cambridge Scientific Abstracts* also maintains *Environmental Sciences* and *Pollution Abstracts,* as well as databases covering aquatic sciences and agriculture. These databases are available in a variety of formats. If you are using Web technology, you'll find that the databases connect to selected Web sites as well as to a list of references, along with abstracts.

Medline is an indispensable tool for locating medical, veterinary, dental, and other health-related literature. Created by the National Library of Medicine, *Medline* is available from many vendors. *PubMed* is the National Library of Medicine's search engine for accessing *Medline, Pre-Medline,* and additional sources. Again, provided that your library has made the appropriate arrangements, *PubMed* will provide full-text access to some online journals. *PubMed* also offers a "related articles" feature allowing you to access an additional set of references related to a particular document. This feature is also offered by other databases such as *Web of Science.*

Another comprehensive medical database is EMBASE. Available in a variety of formats, EMBASE indexes international literature on human medicine and related subjects such as paramedical professions, drugs, optometry, public health, gene technology, hospital management, and other medical specialties.

The *Cumulative Index to Nursing and Allied Health Literature* (CINAHL) is a major access tool for sources on nursing and allied health professions. Subjects range from respiratory therapy and cardiopulmonary technology to health education, medical records, and many others.

Chemical Abstracts is the world's largest database for accessing chemical literature. It is extremely helpful for locating research in the areas of biochemistry, genetics, environmental chemistry, pollution, and biomedicine.

GEOREF is the key to the literature of earth sciences and geology, including paleontology; *Geobase* indexes geology, geography, and ecology. Both databases are available in a variety of formats.

Geology documents are also found in the *GPO Monthly Catalog* (published by the Government Printing Office), which indexes the vast array of items generated by federal agencies. Biologists may be particularly interested in documents about agriculture, energy, the environment, and health, as well as hearings concerning relevant biological issues. Of course, most government agencies have their own Web presence with internal search engines allowing biologists to locate data sets as well as documents.

INSPEC covers topics in physics, electrical engineering, electronics, telecommunications, computers, control technology, and information technology. However, since there is much overlap in the areas of biology and medicine, INSPEC is also a valuable database for topics in the natural sciences.

Science Citation Index, created by ISI, is an extensive multidisciplinary science database. Coverage and numbers of records are far more extensive than those of *General Science Abstracts*. Available in print, online or on CD-ROM (SCISEARCH), and on the Web (*Web of Science*), this unique database allows you to identify articles that cite a particular author or a particular document. Included in each record is the article's Literature Cited section. Like *PubMed, Web of Science* provides a related articles feature.

You will find that *Science Citation Index,* in its various manifestations, is a search tool with multiple uses. If you already know of one paper relevant to your topic, you can use this to locate other, more recent papers by authors who have mentioned (or cited) it. Chances are that those authors have written about the same subject. First, look up the known paper (say it's by J. L. Jones in 1990). You may find one or more references to other authors who have cited Jones (1990) in their own work. Full bibliographic information on these papers will be provided, and if you are using SCISEARCH or *Web of Science,* you may also be given the abstracts. You can then go to the appropriate journals for the papers themselves.

You can use *Science Citation Index* in several other ways. For example, if you suspect that J. L. Jones has published other papers since 1990, you can find them listed under Jones's name. If you know that scientists at

Northwestern Medical School in Chicago are particularly interested in the contraceptive effects of breast-feeding, you can locate other articles originating at this institution. Finally, if you have thought of one or more key words pertaining to your topic, you can use these to locate other relevant sources.

ISI's *Current Contents,* an example of a "current awareness source," accesses recently published papers by reproducing journal tables of contents. *Current Contents* is available in a variety of formats, and the data are embedded in *Web of Science. Current Contents* covers life sciences, social and behavioral sciences, medicine, agriculture, biology, and environmental sciences. Other current awareness sources include *Current Opinion in Cell Biology, Current Opinion in Genetics and Development, Trends in Biochemical Sciences, Evolution,* and *Trends in Neurosciences.* In addition, your library may subscribe to full-text journal services that are available on the Web. Some, like *Ideal,* have their own search engines or are linked with indexes and abstracts.

Do not expect any one database to provide all the references needed to achieve your research goals. And remember that paper indexes are still highly valuable for historical research, since they include older sources that may not appear in online databases. Consult your reference librarian for instruction about the various features of each database or index so that you can use these resources effectively and efficiently.

■ Plan your search vocabulary carefully.

As you begin your literature review, you'll need to build a vocabulary that, although precise, still allows the broadest flexibility in your search. Key words and authors' names are the chief means of accessing information in databases. If you are having little success in researching a particular topic, ask a reference librarian for help in expanding or revising your search terms. Be sure to consult the online thesaurus for the databases you are using. Note that acronyms, such as AIDS, or abbreviations, such as DNA, are often useful search terms. Be prepared for alternative spellings: for example, the British *colour* instead of the American *color* or *behaviour* rather than *behavior.*

Note that truncation features or wild cards are available for most databases; these allow you to search for the singular and plural forms of a term simultaneously or to access terms containing particular roots. As an example, suppose you are interested in estuaries and decide to search BIOSIS Previews. It may seem logical to start with *estuary* as a search term. However, by using the appropriate truncation symbol (for example, *) to make *estuar*,* you may retrieve a larger number of potentially useful references. The root *estuar** will also locate documents pertaining to estuary, estuaries, and estuarine. You can further refine your search by using additional key words in combination with *estuar*.*

■ **Use the Literature Cited section
of relevant primary sources
to find additional references.**

Once you are well established in the primary literature, one of the most efficient ways to expand your list of references is to take advantage of the experience of established researchers. Each time you find a useful paper, study its Literature Cited section carefully. In it you may discover some additional titles that are relevant to your work. The author's comments about these sources in the Introduction or Discussion section of the paper may give you further clues about their contents, perspective, or significance.

USING THE INTERNET

■ **Investigate information resources
on the Internet.**

Using the Internet, you will find a huge variety of resources, including specialized data files; electronic journals and newsletters; text and software archives; and special-interest discussion groups and mailing lists pertaining to seemingly every conceivable subject — from potato growing to camels, from paleoclimatology to bird banding. Internet resources constitute a bewildering array of connections to academic, commercial, governmental, and private users and services; moreover, these resources are constantly changing. To use the Internet or, in particular, Web sources, you must devise a focused search strategy. Therefore, you need enough background knowledge to make a list of important concepts related to your topic. For example, suppose you are interested in primate behavior and have decided that the following general and specific key words are most significant for your search:

primates ⇒ lemurs ⇒ sifakas ⇒ *Propithecus verreauxi*

behavior ⇒ families ⇒ play

How can you best combine these terms for the most efficient and effective search strategy? First, you need to know something about search engines. A search engine is a program that runs on a graphical browser and allows you to use key terms to conduct a search on the Web. Some search engines, such as *Yahoo!* at <http://www.yahoo.com>, search their own carefully selected database. Metasearch engines, such as <http://www.metacrawler.com>, search the databases of other search engines.

Most search engines use Boolean operators: AND, OR, NOT, or NEAR. If you were searching the Web using the terms above, you could limit or increase your end results by using a combination of key words and Boolean operators. For example, you could expand your search using the combination (lemurs OR sifakas) AND play. By "nesting" the key words connected by OR, you are instructing the program to first look for the terms lemurs or sifakas and then look for documents and Web sites that also include the word play with either lemurs or sifakas. Because your nested terms are connected with the operator OR, your search will include Web sources with either lemurs and play or sifakas and play. If you wish to limit your search, you can choose more specific key words, such as *Propithecus verreauxi* (a type of lemur), and combine them with play. In addition, you may use the operator NEAR in many search engines to indicate that two words may be separated but in close proximity. For example, the search phrase lemurs NEAR play might allow you to pick up a Web site entitled "Play Behavior in Lemurs."

Search engines provide valuable tips on specific usage if you click on the Help button. For an excellent nitty-gritty description of search engines and searching techniques, see BARE BONES 101: A Basic Tutorial on Searching the Web: <http://www.sc.edu/beaufort/library/bones.html>.

■ Evaluate Web sources carefully.

Even the most refined search strategy will yield an overwhelming number of Web sites. Of course, you must select the most relevant references to use for your paper. Generally you will need to focus on primary rather than secondary sources. Many print, peer-reviewed journals have migrated to digital counterparts (for example, *Molecular Genetics and Metabolism*). Some primary journals, such as *Evolutionary Ecology Research,* have no print equivalent but appear only online. Original research also appears on many organizations' Web sites, such as that of the Center for Disease Control (CDC) at <http://www.cdc.gov>. The CDC publishes morbidity and mortality statistics, epidemiology reports, and many other kinds of invaluable data.

Many Web sites are clearly secondary material and may provide substantial background information, but you will need to evaluate their relevance in much the same way you would evaluate any document your research uncovers. Questions to ask yourself include the following:

- Who is the author (or group of authors), and where was this research conducted (an academic setting, a government or corporate laboratory)?
- What can you determine about the credentials of the author(s)?
- What other works have the author(s) published?
- Has anyone cited those publications?
- Did a particular organization sponsor the Web site?
- If so, is this organization a not-for-profit or a commercial enterprise?
- Does the organization represent a special interest group (Defenders of Wildlife, American Farm Bureau, and so on)?

- What are the assumptions or arguments presented by the author(s) or organization?
- Are these arguments balanced, or do they reflect bias?
- Are conclusions supported by references?
- Are there links to other Web sites? If so, are these relevant to your topic?
- On what date was this Web site made available? When was it last revised?

The following Web sites may be helpful to you in evaluating Web sources:

Evaluating Web Sites: Criteria and Tools
<http://www.library.cornell.edu/okuref/research/webeval.html>

Evaluating Information Found on the Internet
<http://milton.mse.jhu.edu:8001/research/education/net.html>

Thinking Critically about World Wide Web Resources
<http://www.library.ucla.edu/libraries/college/instruct/web/
    critical.htm>

■ Keep careful records.

Your research will be more efficient, and you'll avoid unnecessary repetition of steps, if you keep a record of the search process as you go along. Write down, for example, which databases you consulted, what key words or authors' names you used, how far back in time your search extended, and so on. Also, record *complete* bibliographic details about all potentially useful sources, including those on the Web. See Chapter 6 for details on what information you should include.

■ Allow ample time.

It is easy to underestimate the amount of time it takes to do a thorough literature search. Choosing a preliminary topic takes time; narrowing and focusing it takes even longer. Successful library research is not simply a matter of accumulating many references as quickly as possible — this will just make you feel overwhelmed. Stop at intervals to read or at least skim your best sources, and use them to help you refine your purpose and scope so that your search becomes increasingly focused and more specific.

Remember, too, that your literature search will probably turn up many promising titles that are not found in your own library. Here you have options. One is to use the interlibrary loan system: at your request, your library can usually borrow a book you need, or obtain a copy of an article from a journal carried by another institution. However, this process may take several days or even weeks, especially in the case of more obscure materials, so you need to plan ahead. Be selective in your requests, since an interlibrary loan service costs time and money. Another option is to write or e-mail authors directly and request reprints of their papers. As with interlibrary loans, allow plenty of time.

READING SCIENTIFIC PAPERS

If you are not used to looking at biological journals, you may find research papers hard to read and understand. Here are some guidelines to help you get the most out of a scientific paper.

■ Acquire some background knowledge.

Papers reporting original research are written for a relatively specialized scientific audience, and thus assume some knowledge of the subject matter and the vocabulary. You can't, of course, become an expert in any field overnight. However, you can acquire some familiarity with the major ideas and key terms before tackling the primary literature. If you build this foundation, you will be less likely to feel overwhelmed later. See page 6 for suggestions about locating general, introductory references. Also useful in this regard are scientific dictionaries and encyclopedias.

■ Read the Abstract first.

The Abstract—a brief summary of the paper—will give you an overview of the study and help you decide whether to read the rest of the paper. Don't feel intimidated by abstracts containing unfamiliar terms or ideas. Often the Introduction and Discussion sections will provide much of the explanatory material you will need for a fuller understanding. Your main task, at this point, is to decide whether the paper might be useful to you. Note that, in some papers, the main text is followed by a Summary section rather than preceded by an Abstract.

■ Understand the basic aims of the study.

If the paper seems relevant, read the Introduction carefully. *Why* did the author(s) conduct the research? What were the major hypotheses or predictions? Authors generally end the Introduction with a brief statement of their objectives; some also include a succinct statement of the major findings.

■ Don't get bogged down in the Materials and Methods.

It is easy to become overwhelmed by the specialized terminology found in many Materials and Methods sections. However, unless your research requires close examination of the methodology, you need not understand every procedural detail. Instead, work at grasping the author's basic approach. Try summarizing the methods in a few sentences, using your own words. Make sure you understand how the experimental design, use of controls, sampling techniques, or other methods relate to the specific objectives of the study. Once you have a sense of the major findings and conclusions, you can reread the Materials and Methods section, if necessary, to clarify selected points.

■ Read the Results carefully, focusing on the main points.

Do not panic if you don't understand the quantitative details. Focus first on the major qualitative findings, and on the author's verbal summaries of quantitative data. These will lay the groundwork for understanding the more difficult material. The author may sum up main points in topic sentences (see p. 132) at the beginnings of paragraphs. Pay attention to comments in the text about figures or tables; they will help you pick out important quantitative trends or relationships. See Chapter 3 for a description of the common types of graphs used in biology and the sorts of information each is designed to convey.

■ Understand the author's argument.

Pay particular attention to the Discussion section, for it is here that the author attempts to tie together the various components of the study. Do the data support the author's conclusions, and if so, how? What does the author suggest is the major contribution of this study? What questions remain for further research? How can you use this paper in your own work?

■ Plan on rereading important papers, possibly several times.

Many of your sources may be too detailed or too difficult to understand in a single reading. Plan on returning to them once you have a firmer background and can better appreciate the specific contributions of each paper.

Also remember when working with primary literature that it is easy to get sidetracked by the specifics of a particular study, thereby losing sight of the broader picture. Keep your own research in mind. Get a general grasp of each author's research; then focus on whatever aspects are relevant to *your* objectives. The importance of a particular article may not be immediately apparent. You may need to skim through many papers at first to get your bearings and return later to those most central to your topic.

TAKING NOTES

■ Avoid plagiarism: take notes in your own words.

Plagiarism is the theft of someone else's words, work, or ideas. It includes such acts as (1) turning in a friend's paper and saying it is yours; (2) using another person's data or ideas without acknowledgment; (3) copying an author's exact words and putting them in your paper without quotation marks; and (4) using wording that is very similar to that of the original source, but passing it off as entirely your own even while acknowledging the source.

This last example of plagiarism is probably the most common one in student writing. Here is an example.

ORIGINAL PASSAGE	A very virulent isolate of *Alternaria mali,* the incitant of apple blotch, was found to produce two major host-specific toxins (HSTs) and five minor ones in liquid culture. The minor toxins were less active than the major ones, but were still specifically toxic to the plants which are susceptible to the pathogen. (Kohmoto, Kahn, and others 1976, p 141)
PLAGIARIZED PASSAGE	Kohmoto, Kahn, and others (1976) found that a very virulent isolate of *Alternaria mali,* the incitant of apple blotch disease, produced two main host-specific toxins, as well as five minor ones in liquid culture. Although the minor toxins were less active than the major ones, they were still specifically toxic to the susceptible plants.

Although the writer has altered a few words here and there, the second passage is strikingly similar to the original. *It is still plagiarism if you use an author's key phrases or sentence structure in a way that implies they are your own, even if you cite the source.* The only way to make this passage "legal" as it now stands is to enclose everything retained from the original wording in quotation marks. Better yet, you should first determine which facts or ideas in a source are relevant for your purposes, and then put these in your own words and word order.

Plagiarism of this kind is usually unintentional, the result of poor note taking and an incomplete understanding of the ethics of research and writing. Typically the problem arises when you lean heavily on notes that consist of undigested passages copied or half-copied from the original source. These become the source of all the information and ideas for your paper. When you sit down to write the first draft, it is all too easy for this material to end up barely changed as the backbone of your paper. Thus your text becomes an amalgamation of other people's words disguised as your own. Even if you cite references for the facts and ideas, you are still guilty of plagiarism because the wording is not completely yours.

Another problem with this kind of note taking is that it consists of reading without thinking. It allows you to speed through a stack of references without necessarily understanding the material. It conflicts with your major purpose in conducting a literature review: to evaluate and interpret information on a subject. You need to start making judgments, comparisons, and contrasts while you are still working with the original sources; otherwise, your prose is just a mosaic of other people's material. Your own paper, like professional papers, should be more than just a sum of its parts.

Form the habit of taking notes mainly in your *own* words. If you are not used to doing this, you may be frustrated by the additional time it takes. However, once you start the first draft, these notes will save you much time and effort. You will have already worked through difficult material, weeded out many inconsistencies, responded to the conclusions of other authors, and made connections among related ideas. Much of the preliminary work will have already been done.

To take notes effectively you need to understand how to *paraphrase* and *summarize* material. A paraphrase expresses certain facts or ideas in different wording — your own — but in about the same number of words as the original. A summary expresses the important facts and ideas in fewer words than the original; for example, the abstract of a research paper is a summary. Both paraphrasing and summarizing require that you understand material fully before you write about it. Although you will probably use both methods as you work through your sources, you'll find that learning how to identify and summarize the points that are *most* relevant to your particular needs is a highly valuable research skill. For example, the writer of the plagiarized passage above might have written the following in his or her notes, to be incorporated later in the final paper.

```
Kohmoto, Kahn, and others (1976) cultured the
fungus Alternaria mali, which causes apple
blotch, and isolated seven different toxins. Of
these, two were particularly toxic to suscep-
tible plants.
```

■ Use an orderly system.

A common method is to use index cards, putting one idea or group of related ideas from a single source on each card. The cards thus contain manageable units of information and can be shuffled around at will as you organize your paper. However, such a method can be cumbersome, and many people feel constrained by the small size of the cards. Scientific topics often require longer, more detailed notes that cannot fit on index cards.

If you have a laptop computer, you may want to record notes directly into a computer file. Later you can cut and paste portions of your notes directly into your paper. An alternative method is to take notes on whole sheets of paper, writing on just one side so that you can cut, paste, and arrange notes later as you prepare the first draft.

Obviously, you need not take notes in complete sentences. In fact, if you try to restrict yourself to succinct phrases, you'll be even less likely to reproduce the exact wording of the original. If the author's own words *are* indispensable, enclose them within quotation marks along with the page number of the source. Do this for entire passages you wish to preserve, as well as for important words or phrases mixed in with your own notes:

> *J. concludes that "despite the predictive power and elegance" of the scientific method, it can give us only a "rough approximation" of what the natural world is like (Johnson 1933, p. 4).*

You also need a foolproof method to distinguish between an author's ideas and your own. For example, you might use a yellow marker to highlight your ideas, or put your initials, the word *me,* or some distinctive symbol in front of any speculations and conclusions that are strictly your own:

> *B. suggests that light availability is the most important factor here. (me) What about moisture requirements? Not discussed.*

■ Be selective.

You will waste time and effort if you take copious notes on every source you encounter. Read first; take notes only when you have decided that the reference may be useful. Resist the temptation to photocopy *every* potentially useful source in its entirety; your research will be more productive if you try to make decisions, as you go along, about the relevance of each reference to your own objectives. If an article seems only peripherally connected, or if you are not sure about its relevance, you may wish to photocopy the Abstract or other selected portions, but reserve your photocopying budget for those few papers that seem central to your work and that you may need to reread later.

■ Record full and accurate information about your sources.

Keep a master list of all the references you consulted. Some people do this on whole sheets of paper or in a separate computer file; others list each source separately on an index card, so that when it is time to assemble the Literature Cited (References) section, the cards can be shuffled and arranged in the proper order. Whatever system you use, be sure to separate the sources that you *cited* (referred to) in your paper from those you may have read but did not cite. Only cited sources are typically included in the Literature Cited section of biological papers.

Before you start your search, learn the kinds of bibliographic information you will need to report for each kind of reference you use (see Chapter 6). If you are not sure whether certain details are necessary, write them down anyway. It is easier to omit unneeded material when you eventually type your references than to spend time searching for missing publication dates or page numbers. If you photocopy part or all of an article, remember to write full bibliographic information about the source directly on the photocopy, so that you are never in doubt about its origin. Remember that Web sources need to be fully documented; see Chapter 6 for details.

CHAPTER **2**

Handling Data

Whether in the laboratory or the field, much of your training as a biologist involves learning how to collect, manipulate, and interpret both
quantitative (numerical) and *qualitative* (nonnumerical) data. The usefulness
of these data depends on such factors as your experimental design or sampling procedures, your choice of equipment and skill in using it, the statistical techniques used for analysis, and (in a field study) the environmental
conditions. You cannot foresee all the problems that might detract from
the validity of your results. However, you can anticipate and avoid many
potential difficulties by thinking carefully about the kinds of data you want
to collect and how best to handle them. Thorough coverage of statistics
and experimental design is well beyond the scope of this book; however,
the general guidelines below may help you in the initial stages of your
work. (For more specialized references, see Additional Readings on
p. 193.)

GETTING STARTED

■ Start your research with questions leading to a specific prediction or hypothesis.

This may sound like obvious advice to anyone acquainted with the
scientific method (see Introduction). Nevertheless, beginners sometimes
plunge headlong into research with little sense of purpose. Most experienced biologists do considerable thinking, planning, and preliminary data
collection to get their bearings. Such work may be relatively unfocused
and tentative, but it provides the grounding for the next step: the phrasing
of key scientific questions and the formulation of specific predictions or

hypotheses. Without *direction* and *purpose* in your research, you will simply amass large quantities of information but have little sense of what to do with it.

■ **Understand how your specific study fits into a broader context.**

Without such a theoretical background, you may waste time collecting inadequate data. If you are embarking on an independent research project, locate and read the established literature in your subject area (see Chapter 1). You should plan your own work so that it builds on and complements that of other researchers, rather than merely duplicating well-established findings. In addition, if someone has reported problems with a particular sampling method, experimental procedure, study species, or piece of equipment, you may be able to avoid the same difficulties.

Even in the case of structured labwork for a biology course, you need to appreciate *why* you are doing each activity. Such lab exercises are usually designed to illustrate general principles and to acquaint you with standard scientific procedures. Read your lab manual before coming to class, and try to understand how each experiment or field trip relates to a broader topic covered in lectures or the text. Only then can you collect and manipulate the right kinds of data effectively.

■ **Identify the key variables so that you can plan your method of data analysis.**

Are the important variables in your study *attributes* (qualitative categories such as smooth seeds versus wrinkled seeds, or red eyes versus white eyes)? If so, you will likely be interested in the *frequency* with which individuals fall into each category. Many studies involve *measurement variables* (quantitative characteristics such as weight, height, or temperature). Measurement variables can be *continuous* (taking any value within a given range, such as abdomen length in bumblebees) or *discontinuous* (taking only integer values, such as the number of leaves per zinnia plant). A third class of variables, *ranked variables,* includes ones that are ranked in order of magnitude, as in the case of the order of flowering (first, second, third, and so on; or 1, 2, 3, . . .) among a group of plants.

You will also need to determine what you will consider *independent* as opposed to *dependent* variables in your study. A dependent variable is so named because its value depends on (is affected by) the value of one or more independent variables. In other words, a dependent variable is a *function* of an independent one. For example, the growth rate of a bacterial culture is a function of temperature; transpiration rate in plants is a function of the relative humidity of the surrounding air; and so on.

The way in which you categorize and measure the key variables in your study will play a large role in determining the kinds of statistical analyses you'll need to use later.

■ Decide which variables you will hold constant or control for, either experimentally or statistically.

When investigating differences between two or more samples, or between expected and observed results, you must attempt to minimize the influence of some variables while isolating and evaluating the effects of other variables that are of interest to you. All this is part of careful experimental design — or in the case of descriptive field data, thoughtfully planned observation or sampling techniques.

Suppose, for example, you are studying the effect of pH on a particular enzyme-catalyzed reaction. Here the independent variable is pH; the dependent variable, the reaction rate, is expected to vary with pH conditions (and hence enzyme activity). You plan to manipulate values of pH in the laboratory, and to collect data on reaction rates under these different conditions. However, you assume that temperature will also influence the rate of the reaction; therefore, it will be necessary to *control* for temperature — for example, by conducting the experiment in a water bath under constant temperature conditions.

As a field example, consider a field study of mate-guarding by male dragonflies. You predict that males will spend less time perching and more time guarding their mates under certain conditions — for instance, when the population density is high. However, you suspect that ambient temperature may affect both population density and perching behavior. Although you can't control temperature conditions in the field, you can try to minimize the unwanted influence of temperature in your study by restricting your observations to a narrow, arbitrary range of temperature conditions, say 25–27°C.

As still another example, suppose you are testing the effectiveness of a 0.1% aqueous solution of gibberellic acid in stimulating radicle (embryonic root) elongation in germinating lupine seeds. You prepare 20 samples, each containing 25 seeds that have been soaked in distilled water for 96 hours. The seeds in 10 of these samples are treated with 10 ml of gibberellic acid solution, whereas the seeds in the second, *control* set of 10 samples receive 10 ml of water only. You then subject all samples to the same temperature, light, and humidity conditions by conducting the experiment in a growth chamber. If you find statistically significant differences in radicle length between the treatment and control groups, you can reasonably attribute these differences to the effect of gibberellic acid.

You can also *statistically* control for a particular "nuisance" variable, although this procedure should never substitute for careful use of experimental controls. For example, partial correlation analysis allows you to study the association between two dependent variables of interest while controlling for the effects of a third. Analysis of covariance is appropriate for examining the relationship between a dependent and an independent variable while controlling for the effects of a second independent variable. Check with your instructor for guidance on these and other experimental procedures. (See also the references in Additional Readings, p. 193.)

■ **Where appropriate, randomize your method of data collection.**

Where individuals must be allocated to different treatments, or must be selected from a larger group for measurement, you should avoid possible bias by choosing those individuals *randomly*. For example, suppose that in a plant growth study you have 200 bean plants growing in trays and wish to harvest 10 plants each week for dry weight determination. Choosing individuals randomly means that each plant in the tray has an equal chance of being selected; thus there is no danger of subconsciously picking plants of a particular size or appearance that might match the outcome you expect, or hope, to occur. One way to ensure randomization is to number each plant at the beginning of the experiment and then use a table of random numbers to select the 10 plants to be harvested each time. Such tables are found in statistical handbooks; also, many computer statistical applications (and even some calculators) have random number generators that make the task even easier.

RECORDING AND ORGANIZING YOUR FINDINGS

■ **Devise effective data sheets.**

Decide exactly what information you plan to record and in what form to record it. Make up a preliminary data sheet, using headings to create a series of columns in which to enter the values you think you will need.

Figure 1 (p. 24) is a sample data sheet for a laboratory project on the role of phytochrome in seed germination, using five different plant species. Its orderly arrangement allows you to record data quickly and systematically. Space is provided for recording the species, the date (useful for record-keeping purposes), and germination differences among the treatment groups. Two of the column headings already include the unit of measurement (here, percent). Unvarying data — for example, incubation conditions, and exposure times for red and far-red light — are recorded elsewhere in notes for the Materials and Methods section of the lab report.

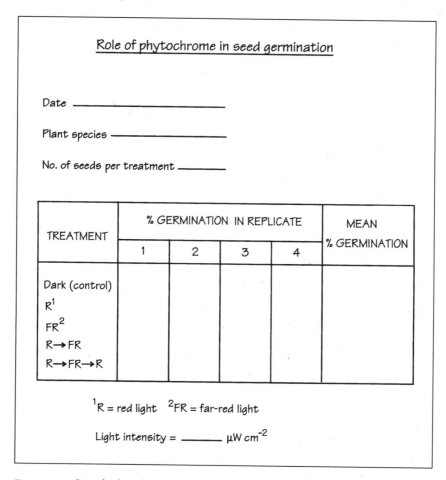

FIGURE 1. Sample data sheet for a laboratory study

For many researchers, spreadsheet software programs have greatly simplified the process of recording and analyzing data. Figure 2, created using Excel (Microsoft Corp.), is a computer-generated version of the data sheet in Figure 1. Other competing products, such as Corel Quattro Pro (Corel Corp.) and 1-2-3 (Lotus Development Corp.), work similarly to Excel. Note that, as in Figure 1, columns in Figure 2 are set aside for raw data pertaining to the different treatment groups. Additional columns (F and G) contain formulas set up in advance by the researcher for calculating particular descriptive statistics (here, means and standard deviations). As data are entered into columns B, C, D, and E, these statistics are automatically calculated and recorded. The accompanying graph is dynamically linked to the data table: the vertical bars are plotted from the values in column F, and the error bars from those in column G. The graph is updated

Mean and standard deviation values are calculated by formulas and are automatically updated as new entries are made in column E, or as changes are made in B, C, or D.

Values for replicate 4 remain to be entered in column E.

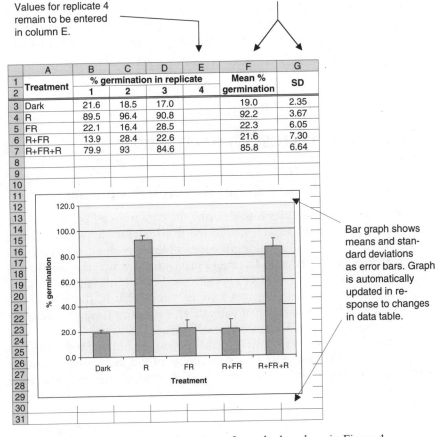

	A	B	C	D	E	F	G
1	Treatment	% germination in replicate				Mean % germination	SD
2		1	2	3	4		
3	Dark	21.6	18.5	17.0		19.0	2.35
4	R	89.5	96.4	90.8		92.2	3.67
5	FR	22.1	16.4	28.5		22.3	6.05
6	R+FR	13.9	28.4	22.6		21.6	7.30
7	R+FR+R	79.9	93	84.6		85.8	6.64

Bar graph shows means and standard deviations as error bars. Graph is automatically updated in response to changes in data table.

FIGURE 2. Computer-generated version of sample data sheet in Figure 1

each time data are entered; it can also be positioned anywhere on the spreadsheet and can be resized and edited at will.

Remember that a good data sheet is the product of much time and thought. Figuring out appropriate headings for all the columns forces you to evaluate the effectiveness of your methods and the rationale for collecting each set of information. However, even the most carefully planned data sheet may prove unsatisfactory the first time you use it, because methods or formats that seem feasible on paper do not always work in practice. You may need to modify your original sheet to account for problems you had not anticipated, or for additional variables that need to be considered.

Here are a few more suggestions about recording data:

1. If you are recording data by hand, use pencil or waterproof ink, especially for fieldwork. After each laboratory or field session, look over the data sheets carefully. Check that your writing is legible. Record additional comments or questions immediately, before you forget them.
2. When transferring information to a computer data file, double-check your data for copy errors. Take the precaution of keeping backup copies of all files, and store these in a different place from the originals.
3. Keep your raw data sheets; you may need to look at them again later to check an anomalous result, or to reexamine your data in light of subsequent research.

USING STATISTICS TO ANALYZE YOUR DATA

■ Make appropriate use of descriptive statistics.

A major use of statistics is to describe the data from a particular *sample* (for instance, 117 seed capsules collected from 63 *Linaria vulgaris* plants), generally with the aim of making broader statements about the *population* (here, the seed capsules of *all L. vulgaris* plants) from which the sample was drawn. In biology, the most commonly used descriptive measure is the *mean* ($\bar{x}$), or arithmetic average of a set of values (that is, the sum of the values divided by the number of values). This measure is usually what we are referring to when we talk about the "average."

Another commonly used descriptive measure in biology is the *standard deviation,* which gives a sense of the variability of the data within a sample. The standard deviation is computed by first determining the difference between the sample mean and each measurement value. These differences are then squared and summed, and the resulting sum is divided by the number of values minus one. The end result of these operations is a statistic called the *variance;* its square root is the standard deviation. Two sets of data may have identical means but different standard deviations. For example, consider the following two samples: (a) 21, 21, 22, 22, 22, 22, 23, 23, and (b) 3, 9, 14, 19, 24, 29, 37, 41. Each data set consists of eight values; in both cases, the mean is 22.00. However, the measurements in the first sample show little variability; here the standard deviation is 0.76. By contrast, those in the second sample vary considerably, and the standard deviation is much greater (13.32). Because mean values on their own convey only limited information, they are usually reported along with the sample size (number of measurements), as well as the standard deviation or some other measure of variability. Where consideration of the highest and lowest values is important, the *range* may be given instead of, or in addition to, the standard deviation. Here, the range (defined as the maximum *minus* the minimum value) for sample (a) is only 2, whereas that for sample (b) is 38.

Certain measures derived from the standard deviation — for example, the *standard error* ($s_{\bar{x}}$) and 95% *confidence intervals* — are often used to indicate the reliability of the sample mean as an estimator of the true mean of the population. There is a 68.26% chance that the interval composed of $\bar{x} \pm s_{\bar{x}}$ includes the true mean of the population. A 95% confidence interval (derived mathematically from the standard error) is a wider interval with a 95% chance of including the true population mean. Thus, the larger the standard error (and hence the 95% confidence interval), the less reliable is a sample mean as an estimate of the population mean. See Figure 8 in Chapter 3 for a graphical depiction of standard errors.

Be sure you understand your purposes for reporting *any* descriptive statistics about a set of values. Remember that misuse of statistics can distort data rather than illuminate them. Consult your instructor, or see the Additional Readings on page 193, for guidance in making appropriate use of these and other statistical measures.

■ When using statistical tests, understand the role of a null hypothesis.

A *null hypothesis,* or hypothesis of no difference, assumes that there is no difference among two or more populations from which samples have been drawn, or no difference between results obtained and those expected. For example, your null hypothesis in an insect physiology experiment might be that there is no difference in adult body weights of cockroaches raised on two different diets. In a lab exercise for a genetics course, your null hypothesis might be that the proportion of yellow seeds to green seeds in peas does not differ from the expected ratio of 3:1. Biologists subject data to particular statistical tests in order to determine if they can *reasonably reject* the null hypothesis in favor of an alternative hypothesis.

Suppose, in the lupine study mentioned earlier (p. 22), you find a mean radicle length of 5.8 mm in the seeds treated with gibberellic acid and a mean length of 5.1 mm in the untreated seeds. The null hypothesis assumes that treatment with gibberellic acid has no effect; that is, that mean radicle lengths do not differ between populations of treated and untreated lupine seeds. The alternative hypothesis is that a difference in mean radicle length does in fact exist. Because the mean lengths are numerically different, you might be tempted to conclude that the null hypothesis is false and that the two groups do differ in mean radicle length. You might further propose that this difference is tied to the presence or absence of gibberellic acid. However, there is still some possibility that such a finding might be due purely to chance, and that the null hypothesis is true. This is especially likely if the variability in each sample is large. Doing the appropriate statistical test gives you a measure of the likelihood of wrongly rejecting the null hypothesis.

In this case, you can perform either a *t*-test or an analysis of variance (ANOVA). A *t*-test is appropriate only for a two-sample case; ANOVA is more generally applicable to comparisons of two or more samples. In the present, two-sample case, the two tests are equivalent. Each of these methods generates a test statistic that you compare with a critical value in a table to determine the probability (*P*) of getting a difference between the sample means as large as you observe (or larger) by chance, if the null hypothesis is, in fact, true. Biologists generally feel confident about rejecting the null hypothesis if this probability is less than 5% (expressed as $P < 0.05$). Results associated with $P < 0.05$ are generally considered *statistically significant*. Of course, rejection of the null hypothesis does not *prove* it to be false; rather, it suggests that the probability of wrongly rejecting the null hypothesis, on the basis of your experimental data, is sufficiently small as to make this error unlikely.

■ Make sure your data meet the assumptions of the statistical tests you use.

Understand the rationale for using the particular analysis you choose for your data. Some statistical tests, such as chi-square, are used for frequency data. Many others, collectively called *parametric* tests, are designed for measurement variables. Some examples of parametric tests are analysis of variance, *t*-tests, and Pearson correlation. Do not automatically assume that a parametric test is appropriate for your analysis. All parametric tests make certain assumptions about the samples in relation to the populations from which they are drawn. Specifically, for each sample, these tests assume that quantities called *residuals* (each being a sample value minus the group mean) have equal variance and are *normally distributed*. (When values in a normal distribution are plotted according to their frequency, we get the familiar bell-shaped curve, which is generated by a precise mathematical equation.) Before conducting any parametric analyses, you should check how well your data fit the above assumptions. Computer statistical applications, such as those listed below, include various tests that serve this purpose.

When the assumptions of parametric analyses are not met, you have two options. One is to transform the original values to logarithms, square roots, squares, or reciprocals. Just remember that the transformed values must meet the assumptions of parametric tests. The second option is to use one of a variety of nonparametric tests. Such tests do not assume normal population distributions. However, they do require, among other things, that the distributions have a similar shape. Originally devised for ranked variables, nonparametric tests can be used for measurement variables as well, but they are generally less powerful than parametric analyses if the assumptions of the latter are met. Examples of nonparametric tests include the Kruskal-Wallis test, Mann–Whitney U-test, and Spearman rank correlation. Further information on parametric and nonparametric tests can be found in any general statistics text.

■ Investigate the statistical software applications available at your institution.

Computer software applications have greatly simplified data analysis, making even highly complex statistical procedures accessible to novices. For example, SYSTAT (SPSS, Inc.), available in Windows and Macintosh versions, does almost every imaginable statistical test as well as graphics. Student SYSTAT is a more affordable but still highly useful version of SYSTAT that is accompanied by a handbook on statistical procedures. JMP (SAS Institute, Inc.) is a user-friendly and full-featured statistical application widely used on both Windows and Macintosh computers. BMDP (BMDP Statistical Software), SPSS (SPSS, Inc.), and SAS (SAS Institute, Inc.) were originally developed for mainframe computers, but are now widely used on personal computers. Certain applications specialize in producing high-quality graphs suitable for publication. For example, most of the figures in Chapter 3 were prepared using SigmaPlot (SPSS, Inc.).

It is well worth the time and effort to determine what statistical applications are available through the computer facilities of your institution. You might also consider purchasing some software of your own. Remember, however, that although it is tempting to plug your data into any one of a number of statistical tests, many of them may be inappropriate. The responsibility for selecting the correct procedure still rests with you.

CHAPTER 3

Using Tables and Figures

The text of scientific papers is often supplemented with tables and figures (graphs, drawings, or photographs). Such materials can convey certain types of information much more effectively than words alone. A table can help you compare the results of a variety of chemical analyses. A graph can illustrate the effect of temperature on the growth of bean seedlings. A line drawing can depict an aggressive interaction between two fish, and a photograph can record important features of your study site.

It is not true, however, that tables and figures are essential in a scientific report. Biology students sometimes think that *all* data must be tabled or graphed to produce "professional" results. Even in published papers, unnecessary tables or figures sometimes find their way into print, wasting the reader's time and raising printing costs. Your credibility as a scientist will suffer if you use tables or figures just to impress the reader, when simple text will do as well or better.

A crucial task in scientific writing is deciding on an effective way to display whatever relationships, patterns, and trends are present in the data. You need a good sense of what quantitative data you are seeking and why. What statistical analyses are most appropriate to address the question you have asked? This concern should have high priority at the beginning of your research, not at the end when there is no time to compensate for mistakes, problems, or oversights.

A second and related concern is how to incorporate your quantitative data smoothly into the manuscript. Can your measurements and statistical analyses be summarized easily in the text, or is a table called for? Would a graph be more effective than a table? Rarely is there only one way to depict a given data set; however, one way may be superior to others. You need to review the scientific argument you are making and determine how to display the evidence most convincingly.

TABLES

■ Use a table to present many numerical values or (occasionally) to summarize or emphasize verbal material.

A large amount of quantitative information can be tedious and cumbersome to report in the text. If you put it in a table, the reader can take in everything at a glance, as well as compare one item with another (see Table 1).

Do *not* use a table when it is more important to show a pattern or trend in the data; consider a graph instead. Also, never use a table (or figure) when you could just as easily summarize the same material in the text. Look, for example, at Table 2. This table is unnecessary because it does not contain much information. Its contents can easily be put into words: "In Morrisville, Ohio, I found *Plantago major* at three out of five vacant lots sampled (Cranston Road, Poolville Road, and Hamilton Street; but not at Main Street or Maple Avenue)."

Similarly, Table 3 is unnecessary. The main message (that seeds neither imbibed water nor germinated following any of the treatments) could easily be stated in the text.

Table 4 has not progressed very far from the raw-data stage. The results will be more meaningful after some simple calculations, allowing us to replace the whole array of numbers with two succinct sentences in the text:

Table 1. Soil analyses of four farm fields near Malverne, Vermont

Site	Macroelement concentration (g per 1000 g dry soil)					
	Total N	P	K	Ca	Mg	S
Schwab farm	0.1	0.9	12.2	15.1	11.6	0.6
Toomey farm	0.8	0.7	6.6	8.3	5.4	0.5
Charles farm	2.6	1.1	17.1	15.9	13.1	1.1
Hendrick farm	2.1	0.7	18.3	26.8	13.6	0.8

Table 2. Occurrence of *Plantago major* at five vacant
lots in Morrisville, Ohio

Location of site	Present (+) or absent (−)
Cranston Road	+
Poolville Road	+
Main Street	−
Hamilton Street	+
Maple Avenue	−

After being visually isolated from other fish for 2 wk, each of 11 male *Betta splendens* was shown its reflection in a mirror and observed for 30 s. Collectively, the fish responded by approaching their images ($\bar{x}$ = 3.0 times, s = 2.1), biting the mirror ($\bar{x}$ = 1.4 times, s = 1.0), and erecting their gill-covers for an average of 15.5 s (s = 4.4).

Tables need not be filled with numerical values. They may also be used (sparingly and carefully) to summarize numerous points, to summarize a review of the literature on a topic, to compare and contrast related items, or to list examples or details that would be too tedious to spell out sentence by sentence in the text (see Table 5).

Table 3. Effect of germination inhibitors on *Phaseolus vulgaris* seeds[a]

Inhibitor	Concn. (mol L^{-1})	Imbibition time (days)	Germination time (days)	% germination
Actinomycin-D	0.05	0	0	0
Coumarin	0.023	0	0	0
Thiourea	0.015	0	0	0
2,4-dinitro-phenol	0.028	0	0	0

[a] 80 seeds in each treatment. Seeds treated for 2 days at 40°C.

Table 4. Response of male fighting fish
(*Betta splendens*) to their image in a mirror[a]

Fish nr	Duration of gill-cover erection (s)	Nr of approaches	Nr of bites
1	20	6	3
2	13	5	2
3	10	2	0
4	7	1	1
5	15	1	1
6	16	5	2
7	19	3	2
8	20	1	0
9	14	0	0
10	21	5	2
11	16	4	2

[a]Prior to the experiment, fish had been visually isolated from one another for 2 wk. Observation period for each fish was 30 s.

■ Number tables consecutively, and make them understandable on their own.

Give each table a number in the order in which you refer to it in the text (Table 1, 2, and so on). Even if there is only one table (or figure) in the paper, assign it a number.

A table should be able to stand apart from the text and still make sense to the reader; therefore the *title* must adequately describe its contents. Never use vague, general titles like "Field data," "Test results," or "Feeding studies." Instead, be more specific and informative:

Average accumulations of uranium in lichens sampled at Collins Forest Preserve, Montana

Indices of fear in preexperimentally conditioned and nonconditioned mice

Responses of inexperienced adult blue jays to monarch and viceroy butterflies

Table 5. Selected functions of cytokinins in plants

Function	Plant studied	Reference
Production and proliferation of genetic tumors	*Nicotiana tabacum*	Ichikawa and Syono (1991)
Regulation of cell cycle	*Arabidopsis thaliana*	Soni and others (1995)
Stimulation of cell division	*Nicotiana plumbaginifolia, Arabidopsis thaliana*	John (1998) Riou-Kamlichi and others (1999)
Induction of adventitious shoots in plant callus cultures	*Arabidopsis thaliana*	Brandstatter and Kieber (1998)
Delay of leaf senescence	*Nicotiana tabacum*	Ori and others (1999)
Promotion of chloroplast maturation	*Arabidopsis thaliana*	Chory and others (1994)
Promotion of cell expansion in leaves and cotyledons	*Raphanus sativum*	Rayle and others (1982)
Promotion of lateral bud development	*Arabidopsis thaliana, Nicotiana tabacum*	Medford and others (1989)

The title is usually written as a sentence fragment at the top of the table. (A sentence fragment is a group of words that lacks a subject, a verb, or both, or that does not express a complete thought. Table titles usually lack a verb.) By convention, only the first word of the title is capitalized. If necessary, you may add one or more full sentences after the title to give further explanatory information:

Relative abundance of 11 cyprinid species at five lakes in Nassau County, New York, in 1994 and 1995. Fish were captured by seining on July 7–9 each year.

Footnotes also allow you to add clarifying information to a table. Unlike material appearing after the title, footnotes are less obtrusive and distracting. You can use numerals, lowercase letters, or conventional symbols (such as *) to mark footnotes. Put these as superscripts (such as [a]) at the appropriate places, and list the footnotes in order at the bottom of the table. Do not use more than one kind of designation for footnotes. If those in Table 1 have numerals, for example, use numerals throughout all your tables.

■ Use a logical format.

Arrange similar elements so that they read *vertically,* not horizontally. You can see the rationale for this if you compare Table 6 with Table 7. They display the same data, but Table 7 is more logically set out and therefore easier to read.

Table 6. Growth rate of cell cultures and activity of ornithine decarboxylase (ODC) and succinate dehydrogenase (SDH) in *Pseudomonas aeruginosa* in response to various carbon sources

	Carbon source		
	Glucose	Sucrose	Mannitol
Growth rate (generations/h)	0.93	0.21	0.47
Activity of ODC (μmol CO_2/h)	12.6	6.9	1.5
Activity of SDH (mmol fumarate/h)	137.7	19.3	50.9

Table 7. Growth rate of cell cultures and activity
of ornithine decarboxylase (ODC) and succinate
dehydrogenase (SDH) in *Pseudomonas aeruginosa*
in response to various carbon sources

Carbon source	Growth rate (generations/h)	ODC (μmol CO_2/h)	SDH (mmol fumarate/h)
		Enzyme activity	
Glucose	0.93	12.6	137.7
Sucrose	0.21	6.9	19.3
Mannitol	0.47	1.5	50.9

Arrange the column headings in a logical order, from left to right, and list the data down each column in a logical sequence. List the data neatly under the center of each column heading, and make sure that all dashes or decimal points are aligned (see Table 7). Put a zero before each decimal with a value less than one — for example, 0.23, not .23.

■ Make the contents concise.

A table should display selective, *important* information — not peripheral details, repetitive data, or values that are uniform and unvarying. Unnecessary data will just clutter the table and obscure more relevant data.

For example, if photoperiod, temperature, or other conditions were the same for all lettuce seedlings in your plant growth project, then these are the standard conditions, to be reported in a footnote, the table title, or the Methods section. These data do not belong in the columns of the table. Similarly, some information useful during the research may not be important enough to mention at all; for example, the reference numbers you gave to each of the mice you dissected or to a series of sampling sites. Thus the fact that certain data take up space in your lab notebook does not automatically qualify them for a prominent position in a table.

Abbreviations are used freely in tables to save space and make the contents easier to grasp. In practice you can abbreviate many words in tables that you would not ordinarily abbreviate in the text; for example, temperature (temp), experiment (expt), and concentration (concn). Abbreviations likely to be unfamiliar to the reader (including ones you devise yourself) should be explained in footnotes. No table should contain so many abbreviations that deciphering them becomes an end in itself, demanding unlimited patience from the reader.

Do not put units of measurement after the individual values in a table. To avoid redundancy, put them in the title or the headings (see Table 8, for example). You can also group related column headings under larger headings, so that identical details do not have to be repeated for each column. Row headings can be grouped in a similar fashion.

Make all explanatory material in tables brief and concise. Do not excessively repeat what you have already written in Materials and Methods; instead, give just enough information to refresh the reader's memory and make the table understandable in its context. If the same abbreviations or procedures apply to more than one table, give them only once. In later tables, refer the reader to your earlier footnotes: "Abbreviations as in Table 1" or "Test conditions as described in Table 4."

■ Check tables for internal consistency and agreement with the rest of the paper.

A common fault of student papers is that some tables (or figures) do not make sense in the context of the whole paper. For example, something in one table may conflict with a statement in the text; the data in a table may seem at odds with the trend shown in a graph; or the values in a table may not make numerical sense. Be sure to carefully proofread your tables. Is each value copied accurately from your data sheets? Do all the items in a group add up to the totals shown in the table?

Table 8. Frequency of budding cells and activity of ß-galactosidase in three strains of *Saccharomyces cereviseae* mating type *a* in the presence or absence of α-factor. Data are based on 10 replicates in each treatment

Yeast strain	Mean % cells budding		ß-galactosidase activity (Miller units)	
	+ α-factor	− α-factor	+ α-factor	− α-factor
Wild type	34.0	51.8	872.4	52.0
Mutant 1	51.1	52.6	40.1	31.5
Mutant 2	49.3	43.9	489.1	56.7

FIGURES

■ Use a graph to illustrate an important pattern, trend, or relationship.

A graph is generally superior to a table when you are more concerned about the "shape" of certain data; for example, how activity in a beetle species is related to ambient temperature, or how the concentration of glucose in the bloodstream changes over several hours. Table 9 displays data that become more interesting when we put them in a *line graph* (see Fig. 1), because we get a clearer picture of how this population of plants changes with time.

Note in Figure 1 that the *independent* variable (time) is plotted on the X-axis, and the *dependent* variable (percent of plants surviving at each sampling time) is plotted on the Y-axis. The dependent variable is so named because its value depends on (is affected by) the value of the independent variable. Another way to say this is that the dependent variable is a *function* of the independent one.

Generally, the axes of a graph meet. However, if one or more points fall on one or both axes, it is permissible to shift the axes slightly apart for clarity. This has been done in Figures 1, 5, 6, 7, 8, 11, and 13.

Figures 2, 3, and 4 illustrate other types of graphs that biologists commonly use. Figure 2 is a *bar graph*. The Y-axis depicts a measurement variable (number of flowers per plant), whereas the X-axis depicts qualitative categories, or attributes (plant species). Figure 3 is a *histogram,* or frequency distribution, showing the frequency of observations falling into a series of numerical categories plotted on the X-axis. Figure 4, a *scatterplot,* illustrates the correlation or strength of association between two variables. In this case, the association is between leaf length and width in a particular plant.

Generally, a graph is *not* necessary when trends or relationships are statistically nonsignificant (see p. 70) or when data are so sparse or repetitive that they can be easily incorporated into the text. Figure 5 is an example of a useless graph because the results it portrays can be summarized in a single sentence: "Growth rates of hyphae of *Saprolegnia* and *Achlya* remained constant at 2.38 and 0.92 μm per minute, respectively, before and after treatment with penicillin."

■ Number graphs consecutively, separately from tables, and make them understandable on their own.

Each graph needs to be numbered in the order in which you discuss it. Use a separate series of numbers for graphs and tables — for example, Table 1, Table 2, Table 3; Figure 1, Figure 2, and so on. Use arabic numerals only, not roman numerals. Write out "Figure" in full in the *legend* (caption or title) of each graph. In the text, write "Figure" whenever it appears outside parentheses, but abbreviate it as "Fig." when you put it in

Table 9. Survivorship of *Chaenorrhinum minus* from seedling emergence to production of mature seeds on railroad cinder ballast in Roslyn, New York, 1980

Time (days)	Percent surviving	Time (days)	Percent surviving
0	100.0	56	9.9
7	72.6	63	8.3
14	55.9	70	7.5
21	38.8	77	7.1
28	25.1	84	7.0
35	18.6	91	6.8
42	14.3	98	6.7
49	11.2		

Figure 1. Survivorship of *Chaenorrhinum minus* from seedling emergence to production of mature seeds on railroad cinder ballast in Roslyn, New York, 1980

Figure 2. Production of flowers by three species of
plants in the absence of interspecific competition
(cleared quadrats) and under natural conditions (control quadrats). The plants were *Campanula rapuncu-
loides, Epilobium angustifolium,* and *Hieracium
aurantiacum.* Plotted are means for eight randomly
chosen quadrats, each 1 × 1 m².

parentheses. For example, "As shown in Figure 1, . . ." but "Germination
rates (Fig. 1) . . ."

All figures should make sense apart from the text. The legend should
be specific and informative, not vague as in "Weather conditions" or
"Rainfall." Avoid simply repeating the labels of the two axes (for example, "Amount of rainfall vs time"). Instead, be more descriptive:
"Rainfall fluctuations during the 1980 breeding season."

The legend is usually written as an incomplete sentence with only the
first word capitalized. If necessary, you may follow it with a sentence or
two of additional information, but be brief and concise (see Figs. 5, 8).
Authors submitting manuscripts to journals typically type the figure legends on separate pages, rather than on the figures themselves. For a student paper, you may wish to type or print each legend directly on the figure. (See pp. 157–60 for instructions on manuscript format.)

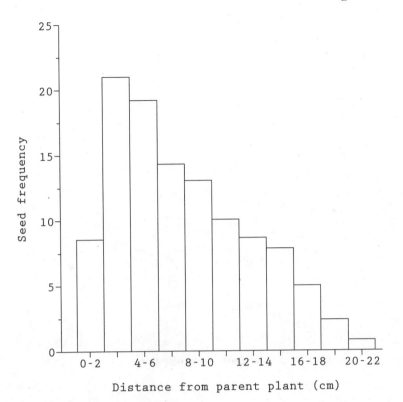

Figure 3. Wind dispersal of seeds from *Chaenorrhinum minus*. Data are from a 2-wk study of a single plant 15 cm tall with 35 mature capsules.

■ Plot data accurately, clearly, and economically.

Figures submitted to professional journals must meet high standards. A sloppily prepared graph will not show off the data to best advantage, and may even cast suspicion on their quality. In student papers, too, accuracy and neatness count. Use standard graph paper, not lined notebook paper, to plot your data accurately. For a neat, professional-looking product, you can then trace this graph onto white bond paper. Also note that computer software applications have greatly simplified the statistical analysis and graphing of data (see p. 29). Check to see what is available at the computer center at your institution.

Remember that the independent variable belongs by convention on the X-axis. Label each axis clearly using large, easy-to-read lettering. Do not forget to specify the units of measurement. Numbers should also be large and legible. You need not number every major interval along an axis; instead, use

Figure 4. The relationship between leaf length and maximum leaf width in *Digitaria sanguinalis.* Data are from 23 randomly selected leaves from five plants, $r = 0.95$.

Figure 5. Growth rate of hyphae of the fungi *Saprolegnia* (filled circles) and *Achlya* (open circles) at 20°C. Penicillin was added to the cultures at 10 min.

index lines to indicate the halfway points between two numbers, as in Figures 1 and 4. This kind of shorthand prevents the graph from being cluttered with unnecessary numbers and gives you more space for essential information.

Many graphs made by beginners waste large amounts of space. Look at Figure 6, for example. The Y-axis is shown originating at zero even though the lowest measurement is 700 cells per ml. And the X-axis is extended far more than is necessary because the last measurement is taken at 60 h.

Figure 7 shows the same data in a more compact format. The Y-axis begins with a value closest to the lowest one plotted, and the X-axis is only as long as needed. If you are used to seeing zero on every scale, this graph may look "wrong." However, indicating the zero mark should never be done at the expense of a graph's clarity and visual effectiveness.

An alternate way of saving space here would have been to start the Y-axis at zero but put a break along it, indicating that unneeded values have been omitted:

You can also save space, as well as compare two or more data sets, by putting them all in a single figure, as in Figure 8. Be sure to label each data set clearly. Explain any symbols you use in the legend or in a key placed inside the graph itself. The latter method makes efficient use of empty space. Do not pack *too* much information into a single figure, or the reader will be unable to make the necessary comparisons. A general rule of thumb is no more than four different symbols and no more than three different lines per graph. However, when many of the data points are close together, confusion and lack of clarity can quickly result, as shown in Figure 9. In this case, the author is better off preparing two separate graphs, one for mayflies and one for stoneflies, and placing one below the other, so that the reader can easily make comparisons (see Fig. 10).

When you plot a series of sample means, it is customary to give the reader a sense of the variability of the data or the reliability of the sample means as estimators of the population means (see p. 68). Figure 8, for example, shows the standard error for each mean as vertical lines extending up and down from each data point. These lines reflect the reliability of each mean as an estimator of the population mean. Using the same

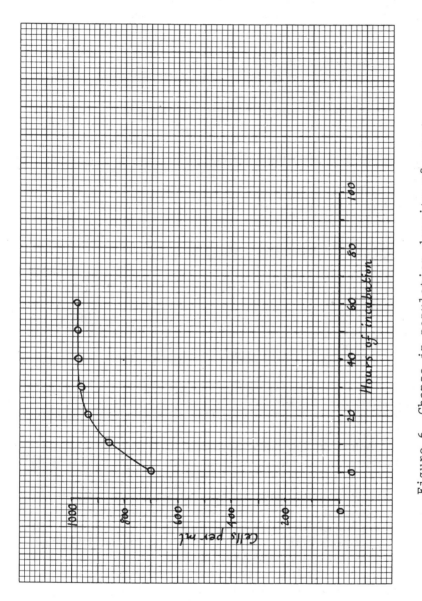

Figure 6. Change in population density of yeast
cells growing at 27°C in malt extract broth.

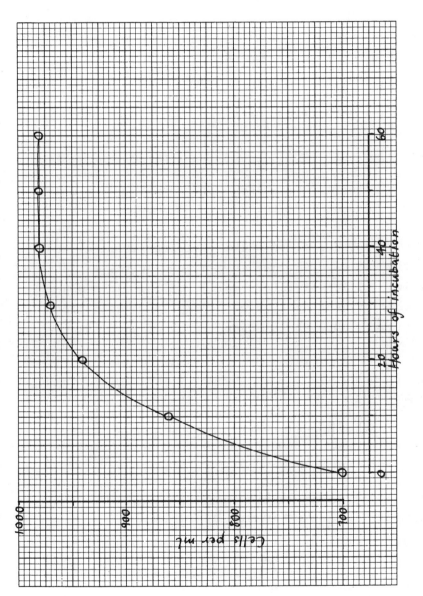

Figure 7. Change in population density of yeast cells growing at 27°C in malt extract broth.

Figure 8. Growth rate of wheat (*Triticum vulgare*) and beans (*Phaseolus vulgaris*) as measured by the ratio of dry weight of leaves to dry weight of the whole plant. Ten plants of each species were sampled at each time. Vertical bars show the standard error of each mean.

method to show the 95% confidence interval for each mean will convey similar information. If you wish to plot the variability of values within each sample, you could plot the standard deviation. If consideration of high and low values is important, you may prefer to show the range of values around each sample mean. Be sure to specify in the legend which statistics you have plotted.

■ Depict data logically, in a manner consistent with your overall hypothesis.

Remember that the data you graph may be subject to more than one interpretation. You can bring out, hide, exaggerate, or misrepresent the message carried by a given set of data by the way you choose to draw the graph. For this reason it is vital that you understand your purpose in constructing the graph in the first place.

For example, does connecting all the dots on a graph always make sense? It depends on your data and how they were obtained and analyzed, as well as on your interpretation. In Figure 11 the data points are *not* connected; instead, the author has drawn a curve by hand to *approximate* the general downward trend suggested by the data. Most of the points are ei-

Figure 9. Abundance of mayflies (*Ephemeroptera*) and stoneflies (*Trichoptera*) at Sacco Lake, Hicksville, Minnesota, in June 1994. Data are based on daily collections at the southern shore of the pond at 1800.

ther touching or very close to this curve. Sampling error (virtually unavoidable in any study) accounts for the fact that all the points do not lie exactly on a smooth curve.

By contrast, in Figure 10 the author *has* joined all the points because his purpose is to show the high day-to-day variability in abundance of mayflies (or stoneflies) at the study site. We assume here that the big gaps between points are not due just to sampling variations; rather, they are meaningful fluctuations in themselves.

In Figure 12 the author has plotted a *statistically derived* (not hand-drawn) line of "best fit" to the data points. Her purpose is to depict a specific mathematical relationship between two variables. The equation for the line, given in the legend, defines this relationship. Figure 13 illustrates a similar situation for a fitted curve. In both figures the authors have

Figure 10. Abundance of (A) mayflies (*Ephemeroptera*) and (B) stoneflies (*Trichoptera*) at Sacco Lake, Hicksville, Minnesota, in June 1994. Data are based on daily collections at the southern shore of the pond at 1800.

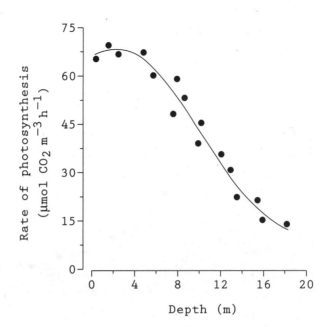

Figure 11. Rate of photosynthesis by marine phyto-
plankton as a function of depth. Data were collected
at Hilton Bay, South Carolina, at 1100, July 31, 2000.

Figure 12. Effect of sodium bicarbonate concentration
on rate of photosynthesis of *Elodea canadensis*, as
measured by production of oxygen bubbles at 18°C. Regression equation is $y = 10.9 + 5.1x$. ($R^2 = 0.79$).

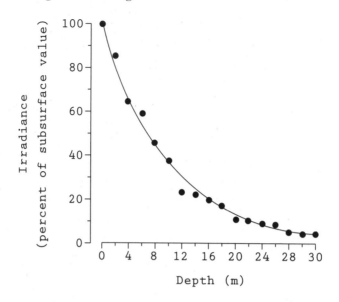

Figure 13. Decrease in light penetration through
water. Data were collected at Glen Cove, Georgia, at
1300, June 22, 1983. Radiant energy (irradiance) was
measured in W cm^{-2} and values at different depths were
calculated as percent of the value just below the
water surface. Regression equation is $y = 101.5e^{-0.1x}$
($R^2 = 0.97$).

assumed that a cause-and-effect relationship exists between the two variables, that is, that values of the independent variable can be used to predict
values of the dependent one. The data points give the reader a sense of the
variability of the observations.

If you look back now to Figure 4 (p. 42), you will notice that the
points are not accompanied by a fitted line. The author's purpose here is
simply to show the interdependence of these two variables, not to suggest
that the values of one are controlled or predicted by the values of the
other. As in other correlation studies (see p. 70), either variable could have
been plotted on either axis, since no assumptions are being made about
which one is dependent and which is independent.

Thus, constructing a graph is more than just an exercise in plotting
points on ruled paper. It requires that you understand the rationale behind
your quantitative methods. How is your interpretation of the data constrained by the statistical analyses you used? What assumptions are you making by describing your data in this particular way? Are these assumptions
valid? Discussion of quantitative methods is beyond our scope here, but the
subject is treated at length in the references listed at the end of this book.

CHAPTER 4

Writing a Research Paper

A research paper is a report of original findings organized into several sections according to a format that reflects the logic of a scientific argument. The paper bears a specific, informative title, usually followed by a one-paragraph Abstract, or summary. First the author states the purpose of the investigation, placing the work in a broader scientific context (Introduction). Then the procedure is described (Materials and Methods), and the findings are presented (Results) and interpreted (Discussion). Finally, the author thanks others for their help (Acknowledgments), and lists the references that were cited in the paper (Literature Cited or References). Biologists follow this format closely when they write a paper for publication. As a biology student, you will use the same organizational method for research projects and laboratory reports.

Many writers feel that the fairly rigid structure of a research paper makes it relatively easy to write. Once you understand what goes where in a research paper, you can break up the writing into discrete tasks and tackle each one in turn. In fact, you do not have to compose the paper in any particular order. Begin with the section that seems easiest; for many people this is the Materials and Methods. The Discussion and Abstract are usually easiest to write last. The Literature Cited section can be assembled as you go along, but you'll need to leave time at the very end to check it for completeness and accuracy.

Of course, the bulk of your writing will have to wait until you have analyzed all the results. However, if you try to visualize each step you take as part of a particular section of the paper, you'll be better prepared to start writing once the time comes. Moreover, you will stay in touch with the aims and progress of your study and be able to modify your approach, if necessary, while there is still time to collect additional data.

One good organizational method is to have a separate file folder, computer file, or notebook section for each section of the paper. While the study is still in progress, jot down procedural details, notes on the strengths and weaknesses of particular experiments, and speculations about the relevance of your findings. Record ideas, questions, and problems as they come to you and file everything in the appropriate place. Do not rely on memory. As good as yours may be, you are bound to forget something unless you write it down. The act of writing will impose order and logic on your activities, especially at the start of the project when you may be inefficient and disorganized. It will also help you crystallize your thoughts, gain perspective, and perhaps see new patterns in the data. You may find that a casual thought captured on paper eventually proves to be a flash of insight.

TITLE

The title identifies the important contents of the paper and orients the reader by specifying the writer's major findings or perspective. For this reason, abstracting and indexing services rely heavily on the titles of papers to organize large quantities of scientific information. Researchers also scan titles to pin down the most relevant references. A vague or inaccurate title can waste a reader's time by suggesting, erroneously, that the paper contains certain information. Even worse, a good paper burdened with a bad title may never catch the eye or the interest of many of its intended readers.

Many writers compose the title last, once they have written at least one draft of the paper and have a better understanding of their purpose and scope. Others devise a working title *before* they do much writing, to help them focus their ideas; then they revise it later, if necessary. See page 158 for instructions on preparing the title page of the final manuscript.

■ Make the title informative and specific.

Organize your title around important words (key words) in the study. Imagine that your paper appears in an index or abstract. How easily could it be located by someone looking up key words pertaining to your subject area? (Biologists submitting papers to journals for publication are often asked to suggest several key words pertaining to their manuscript.)

Your title should indicate the major focus of your study. The following titles, for example, are vague and uninformative compared with the revised versions following them.

| VAGUE | Ecological Studies of Some Northern Lakes |
| SPECIFIC | Seasonal Algal Succession and Cultural Eutrophication in Three North Temperate Lakes |

VAGUE	Dominance Behavior in Rhesus Monkeys
SPECIFIC	Androgen-Induced Social Dominance in Infant Female Rhesus Monkeys
VAGUE	Aspects of Territoriality in Lizards
SPECIFIC	The Adaptive Significance of Territoriality in Iguanid Lizards
VAGUE	Some Changes During Menopause
SPECIFIC	Early Menopausal Changes in Bone Mass and Sex Steroids
VAGUE	A Look at Fungal Toxins
SPECIFIC	The Role of Fungal Toxins in Plant Disease

■ Be concise.

Make every word count. Omit unneeded "empty" words at the beginning of the title.

WORDY	Preliminary Studies on Primary Productivity and Phytoplankton Diversity in Four Wisconsin Lakes
CONCISE	Primary Productivity and . . .
WORDY	Notes on the Effect of Ambient Temperature and Metabolic Rate on the Activity of New Mexican Lizards
CONCISE	Effect of Ambient . . .

Replace wordy phrases with shorter phrases or single words.

WORDY	Why Felids Copulate So Much: One Possible Model for the Evolution of Copulation Frequency in the Felidae
CONCISE	An Evolutionary Model for High Copulation Frequency in the Felidae
WORDY	Studies on the Reproductive Biology of *Drosophila,* Including Sperm Transfer, Sperm Storage, and Sperm Utilization
CONCISE	Sperm Transfer, Storage, and Utilization in *Drosophila*

■ Include appropriate taxonomic information.

If your work features a particular species or larger taxonomic group, specify this clearly in the title. Species should be described by their full Latin names — both genus and specific epithet. Include other information, if necessary, to orient the reader. Such information may include common names (if they exist) or the name of the family, order, or other important group to which the species belongs. Note how the following titles are more informative in the revised versions.

VAGUE	Mating Frequency in Butterflies
VAGUE	Mating Frequency in *Papilio*
SPECIFIC	Mating Frequency in Butterflies of the genus *Papilio*

The first title is too general, unless the writer is discussing *all* butterflies; the second will be confusing to readers unfamiliar with insect (and especially butterfly) taxonomy.

VAGUE	Effect of Hormones and Vitamin B on Gametophyte Development in a Moss
SPECIFIC	Effect of Hormones and Vitamin B on Gametophyte Development in the Moss *Pylaisiella selwyni*

The first title conveys the purpose of the study, but we are left wondering *which* moss is being studied.

VAGUE	Pollination, Predation, and Seed Set in *Linaria vulgaris*
SPECIFIC	Pollination, Predation, and Seed Set in *Linaria vulgaris* (Scrophulariaceae)

Even botanically inclined readers may not be familiar with this particular species. Including the family (Scrophulariaceae) after the scientific name provides the useful information that this plant is in the snapdragon family.

VAGUE	Paternity Assurance in the Mating Behavior of *Abedus herberti*
SPECIFIC	Paternity Assurance in the Mating Behavior of a Giant Water Bug, *Abedus herberti* (Heteroptera: Belostomatidae)

As in the preceding example, the first title of this pair is not informative enough for most readers. Is this organism a mammal? A fish? A bird? The revised version tells us that it is an insect belonging to the order Heteroptera and family Belostomatidae.

Note that there are some species or larger groups whose scientific names do *not* need to be specified in the title, although it is not incorrect to do so. These organisms include those whose common names are familiar, well established, and even standardized (such as American birds), and organisms that are well studied or familiar to most scientists, such as cats or rats. However, it is still necessary to include precise taxonomic information in the text of the paper.

Finally, some species — for example, *E. coli* and *Drosophila* — are well known by their *scientific* names; they can appear without further explanation in the title.

■ Avoid specialized terminology, coined words, and most abbreviations.

As in a published paper, your title should be meaningful on its own; otherwise it will confuse or discourage potential readers. Write the title with your audience in mind. Use only those terms likely to be familiar to most readers — for example, your instructor and your classmates. Avoid esoteric or highly technical language, and do not use abbreviations except for widely recognized ones (such as DNA, RNA, or ATP).

ABSTRACT

■ Summarize the major points of the paper.

The Abstract is a short passage (usually 250 words or less) that appears just after the title and author(s) and summarizes the major elements of the paper: objectives, methods, results, and conclusions. It is typically written as a single paragraph. In the case of published works, a good Abstract helps researchers assess the relevance of a paper to their own research and thus decide whether or not they should read it completely. Scanning the Abstracts of papers is also one way that scientists, when pressed for time, keep abreast of recent literature.

Although readers see the Abstract first, it is easiest to write it last, once you have a good overview of the paper. One way to write an Abstract is to list, one by one, all the important points covered in each section of the paper. Write complete sentences if you can. Make successive revisions, paring the list down bit by bit, omitting peripheral topics and details, until you have revealed the "skeleton" of the study.

■ Be specific and concise.

As a summary, an Abstract must be both informative and brief. Avoid general, descriptive statements that merely hint at your results or serve only as a rough table of contents. Consider every sentence, every word. Could you say the same thing more economically?

The following Abstract from a published paper illustrates how the essential points of a study can be summarized in a single concise paragraph:

> The action of melanin-concentrating hormone (MCH) on melanophores was studied in 27 teleost species. MCH caused melanosome aggregation in all teleosts studied, including two siluroid catfish in which melanin-aggregating nerves are known to

be cholinergic. In most fish, the minimal effective concentration of MCH was estimated to be 10 μM, while in three swellfish examined, it was higher than 10 μM. The mode of action of the peptide was identical in either adrenergically or cholinergically innervated melanophores. It may act through specific receptors on the melanophore membrane. These results suggest that MCH may be a biologically active hormone common to teleosts.

(Nagal, Oshima, Fujii 1986, p 360)

The Abstract below is too general: the student writer has left out important information and refers only vaguely to what was done, how, and why.

This report is a study of the algae in Lebanon Reservoir, using a variety of collecting techniques. Over the study period, there was a gradual replacement of Cyanophyceae by Bacillariophyceae as a result of certain factors, such as temperature. Planktonic communities differed from attached communities because of morphological and physiological characteristics of the algal species in each community. There were marked differences in algal composition between the epilimnion and hypolimnion.

This Abstract could be improved if the writer used more specific language. As it stands, we are left with unanswered questions. What kinds of collecting techniques were used? Were there other important factors besides temperature? What "morphological and physiological characteristics" did the writer study? And so on. We also do not know where Lebanon Reservoir is or why the author did the study in the first place.

The Abstract below is unnecessarily wordy:

This paper reports studies on the territorial behavior patterns shown by males of the dragonfly species *Plathemis lydia* at a small pond in the vicinity of Earlville, New York. A total of 51 male dragonflies were marked with small spots of enamel paint applied to the abdominal portion of the body and were observed under natural conditions during the month of June 1987. It was found that males showed a strong tendency to defend individual areas at the pond. These males

chased away other males of their species, and
also, on occasion, even males of other species
of dragonflies. Threat behavior, as opposed to
behavior involving physical contact, was the
most common aggressive behavior displayed by
males defending their areas against intruders.
Males typically remained at the same area of the
pond for two hours or longer on a single day.
Also, they usually were observed returning to
the same area on successive days. These observations suggest the possibility that male territorial behavior in this species serves the
adaptive function of enabling the owners to monopolize particular areas. These areas may be
visited by females who are in the act of seeking
mates.

This Abstract used 192 words to describe what the same Abstract, written
more concisely, can express in 104 words:

I studied the territorial behavior of male dragonflies (*Plathemis lydia*) at a small pond near
Earlville, New York. Fifty-one males were marked
with enamel paint on the abdomen and observed
under natural conditions during June 1987. Males
defended individual areas from male conspe-
cifics, and occasionally from males of other
species. Aggressive interactions generally involved threat behavior rather than physical contact. Territory owners typically remained at the
same area for at least two hours and returned to
the same location on successive days. Territorial behavior in this species may be adaptive
for males, enabling them to monopolize areas
likely to be visited by females seeking mates.

In the revised version, the author has replaced long, wordy constructions with shorter, more economical wording: for example, "the territorial
behavior patterns shown by males of the dragonfly species *Plathemis lydia*"
has become "the territorial behavior of male dragonflies (*Plathemis lydia*)."

Many unnecessary words ("a total of," "it was found that," "these observations suggest the possibility that") have been omitted. The resulting abstract conveys the same information much more effectively. (See pp. 143–46 for a more complete discussion of wordiness.)

The Abstract below is too long because the author included more information than was necessary. What parts should be left out?

```
The algal succession occurring with autumn over-
turn was studied in the phytoplankton community
of a hard water pond near Madison, New Jersey.
The pond can be sampled easily from all sides
and offers excellent opportunities for ecologi-
cal research. This study was done as a field
project for Biology 314. The dominant species
in October 1986 were Dinobryon sertularia,
Scenedesmus quadricauda, Cryptomonas erosa, and
Cyclotella michiganiana. The first three species
showed significant declines in December, after
fall overturn, whereas the last species became
much more prominent.
```

We do not need to know that the pond is a good place to do research, or that the study was done for a particular biology course.

■ Make sure the Abstract can stand on its own and still make sense to the reader.

Like the title, the Abstract of a published work will be read by many people who may not read the text of the paper. These include researchers using references such as *Biological Abstracts* to locate relevant references on a topic. Therefore, the Abstract must be independent: readers should be able to understand it without being familiar with the details of the study. Omit abbreviations (except for widely known ones such as DNA), and use only those technical terms likely to be familiar to your audience. If certain abbreviations or specialized terminology are absolutely necessary, then explain them. Do not refer to materials that may be inaccessible to your readers, such as your tables and figures, and avoid references to other literature, if possible.

The following Abstract, taken from a student research report, is difficult to understand without having first read the paper:

```
This paper describes a behavioral and computer
simulation study of cognitive differences be-
tween five-year-old boys and girls presented
```

```
with tasks of either Group I, II, III, or IV
difficulty level. We followed a modified version
of the method developed by Wolf (1955, 1977).
When compared with boys, girls scored higher on
what we designated "prepared tests," whereas
boys scored higher on "stimulus-response tests"
(Fig. 3). We concluded that . . .
```

The author of this Abstract provides little information we can understand without having first read the paper. We have no way of knowing what "Group I, II, III, or IV" stands for, what Wolf's method was, what the author's definitions of "prepared" and "stimulus-response tests" were, or what Figure 3 is all about.

Now look again at the Abstract by Nagal, Oshima, and Fujii (1986) (pp. 55–56). Notice that the abbreviation MCH is explained and the language is straightforward. The few technical terms that are used (for example, "melanophores" and "cholinergic") are likely to be familiar to most readers in this field.

INTRODUCTION

The Introduction of a research paper sets the stage for your scientific argument. It places your work in a broad theoretical context and gives readers enough information to appreciate your objectives. A good Introduction "hooks" its readers, interesting them in the study and its potential significance. Thus you, as a writer, must have a firm grasp of the aims, principal findings, and relevance of your research. You may find that the Introduction is easiest to write *after* you have drafted the Materials and Methods, Results, and Discussion sections and have a clearer understanding of just what you are introducing.

■ Orient the reader by summarizing pertinent literature in your field.

An effective way to organize the Introduction is to proceed from the general to the specific, starting with a review of current knowledge about the topic and narrowing down to your specific research problem. Introduce key concepts, define specialized terms, and explain important hypotheses or controversies. As you do so, *document* your writing by citing key references in the field — more general ones first, followed by studies closest to your own research. (See Chapter 6 for instruction on how to document sources.) In this way you can sketch out a framework for the study, orient your readers, and prepare them for what is to follow.

Do not make the Introduction *too* broad or *too* detailed. This is not the place to show off your knowledge about the subject, list every available reference, or repeat material found in any elementary text. Most published papers have short Introductions (often only a few paragraphs) because the writer is addressing readers with backgrounds similar to his or her own. It wastes journal space and the reader's time to give an exhaustive literature review. Similarly, in a paper for a course, write for your classmates and your instructor — people with at least a beginning knowledge of the subject. Discuss only the most relevant concepts and references, and get quickly to the point of the paper.

■ Explain the rationale for the study and your major objectives.

After explaining the broad theoretical context, you are ready to clarify the special contribution your own study makes. How does your work fit in with that of other researchers? What special problem does your study address? What *new* information have you tried to acquire? Why? How? In other words, why (apart from course requirements) are you writing the paper in the first place? Most authors end the Introduction by stating the *purpose* of the study. For example:

> The purpose of this study was to describe the dominant fungi associated with decomposing leaf litter in a small woodland stream.

> In this paper, I report laboratory observations on the effects of crowding on the behavior of juvenile Atlantic salmon (*Salmo salar*).

Remember to clarify the purpose of your paper by first explaining the rationale for the study. What needs to be improved in the following Introduction?

```
The purpose of this project was to study the
algae at two different sites in Payne Creek,
near Stillson, Florida. Chemical and physical
parameters were also considered. The study in-
volved algal distribution in several microhabi-
tats of both stream and marsh environments. This
project was part of a larger class study for Bi-
ology 341.
```

A major problem here is that the writer has not linked the study to a broader conceptual framework. She begins abruptly with a vague statement of her objectives, with no explanation of why they are important. The passage also contains irrelevant information: it is not necessary to state that the study was part of the requirements for a particular course.

By contrast, the writer of this Introduction gives the reader a clear idea of what the objectives are and why they are important:

> It is well known that males of many species of
> dragonflies (order Odonata) guard their mates
> after copulation while the eggs are being laid.
> In some species the male hovers over the female
> and chases away other males; in others, the male
> remains physically attached to the female as she
> moves around the breeding site. Many authors
> have discussed the adaptive significance of
> mate-guarding, particularly with respect to ways
> in which it may increase the reproductive suc-
> cess of the guarder (for example, Alcock 1979,
> 1982; Sherman 1983; Waage 1984). A more complete
> understanding of the evolution of this behavior
> is dependent on detailed studies of many odonate
> species. In this paper, I describe mate-guarding
> in *Sympetrum rubicundulum* and discuss ways in
> which guarding may be adaptive to males.

Some authors end the Introduction by summarizing their major *results* in a sentence or two. This tactic gives readers a preview of the major findings and may better prepare them for the scientific argument that follows. Other writers, along with some journal editors, criticize this practice, arguing that results are already covered in their own section and in the Discussion and Abstract. Ask your instructor what he or she prefers.

MATERIALS AND METHODS

■ Include enough information so that your study could be repeated.

Your methodology provides the context for evaluating the data. How you made your measurements, what controls you used, what variables you did and did not consider — all these things are important in molding your interpretation of the results. The credibility of your scientific argument depends, in part, on how clearly and precisely you have outlined and justified your procedures.

Furthermore, one of the strengths of the scientific method is that results should be reproducible using similar materials and methods. It is not unheard of for a scientist to repeat someone else's experiment and get different results. These conflicting data then point to factors that may have been overlooked, perhaps suggesting different interpretations of the data.

Finally, a complete and detailed Methods section can be enormously helpful to others working in the same field who may need to use similar procedures to address their own scientific problems.

What kinds of information should you include? See the guidelines below.

Materials

1. Give complete taxonomic information about the organisms you used: genus and species names as well as subspecies, strains, and so on, if necessary. Specify how the organisms were obtained, and include other information pertinent to the study, such as age, sex, size, physiological state, or rearing conditions.

> *Cladosporium fulvum* race 4 was obtained from Dr. George Watson, Department of Biology, Colgate University, Hamilton, New York. Stocks were maintained in sterile soil at 4°C and were increased on V8 juice agar at 25°C in the dark. . . .

> Adult American chameleons (*Anolis carolinensis*), purchased from a local supplier, were used for all experiments. They were kept in individual terraria (30 × 30 × 30 cm^3) for at least seven days prior to the start of any study. They were provided with a constant supply of water and were fed crickets, mealworms, and other insects every two days. All chameleons were exposed to 15 h of fluorescent light daily (0800–2300), and air temperatures were kept at 30°C. . . .

2. If you used human subjects, give their age, sex, or other pertinent characteristics. Biologists submitting papers for publication may need to demonstrate that subjects have consented to be involved in the study.

3. Describe your apparatus, tools, sampling devices, growth chambers, animal cages, or other equipment. Avoid brand names, unless necessary. If some materials are hard to obtain, specify where you purchased them.

4. Specify the composition, source, and quantities of chemical substances, growth media, test solutions, and so on. Because they are more widely understood, use generic rather than brand names.

> Sodium citrate, sodium pyruvate, and hydroxylamine were obtained from Sigma Chemical Company, St. Louis, Missouri. All chemicals were of reagent grade. . . .

> For the first series of electron microscopy studies, tissue samples were fixed in 2.8% ultrapure glutaraldehyde, 0.57 mol L^{-1} glucose, and 0.10 mol L^{-1} sodium cacodylate. . . .

5. If detailed information about any of the materials is available in a

standard journal, then avoid repetition by referring the reader to this source.

> I used an intermittent water delivery system similar to that described by Lewiston (1950). . . .

> Tryptone-yeast extract broth (Pfau 1960) was used to cultivate bacterial strains. . . .

Methods

1. Describe the procedures in detail. Do not forget crucial details such as temperature conditions, pH, photoperiod, duration of observation periods, sampling dates, and arbitrary criteria used to make particular assessments or measurements. If you used a method that has already been described in a standard journal, you need not repeat all this information in your own paper; just cite the reference.

> Mycelia were prepared using the fixation and embedding procedures described by Khandjian and Turner (1971). . . .

However, if the reference is hard to obtain (for example, *The Barnes County Science Newsletter*), or if you altered someone else's methods, then supply full information about your procedures.

2. For field studies, specify where and when the work was carried out. Describe features of the study site relevant to your research, and include maps, drawings, or photographs where necessary. If published information already exists on the area, cite sources.

> This study was conducted during June and July 1986 at Bog Pond, 3 km northwest of Barrow, West Virginia. The general habitat of this pond has been described elsewhere (Needham 1967; Scott 1981). The pond is permanent and contains floating and emergent vegetation (mostly sedges, rushes, and algae). It has an area of approximately 1.5 ha. . . .

3. Commonly used statistical methods generally need no explanation or citation; just state for what purpose you used them. If you used less familiar or more involved procedures, cite references explaining them in detail and give enough information to make your data meaningful to the reader.

■ Organize your material logically.

Because the Materials and Methods section contains so many important details, it is easy to forget some, particularly because you are so familiar with the subject. It is also easy to let this section become a confusing, rambling conglomeration of details, with little unity or coherence. Organize this section carefully using an outline, a list, or a plan, along with the detailed notes you compiled while doing the research.

A typical approach is to begin with a description of important materials (study species, cell cultures, and so on) and to move on to the methods used to collect and analyze the data. Field studies often start with a description of the study site. Describe your procedures in a logical order, one that corresponds as closely as possible to the order in which you discuss your results. You may also group related methods together.

Remember that the Materials and Methods section is still part of your text and must be readable. Do not let your paragraphs become disorganized collections of choppy sentences, as in the example below:

FAULTY   Golden hamsters (*Mesocricetus auratus*) used for this research were adult males. Temperature conditions were kept at 22–24°C. Animals were fed Purina chow. Hormonal studies were performed on 23 individuals. The photoperiod was 16 h. Animals were housed with littermates of the same sex, and feeding was once each day. All hamsters had been weaned at three weeks.

REVISED   Hormonal studies were performed on 23 adult male golden hamsters (*Mesocricetus auratus*). All had been weaned at three weeks and housed with littermates of the same sex. They were reared under conditions of 22–24°C and a photoperiod of 16 h, and fed Purina chow once daily.

The revised passage is easier to read and understand. Short, choppy sentences have been combined, and related points have been pulled together. The same information is conveyed, but using fewer words and a more organized style.

If the Materials and Methods section is longer than several paragraphs and involves lengthy descriptions of several topics, you may wish to use *subheadings* (perhaps taken directly from your outline) that break the text into clearly labeled sections (for example, Study Area, Sampling Methods, and Data Analysis; or Test Water and Fish, Testing Conditions, Chemical Analyses, and Statistical Analyses; or Plant Material, Morphometry, Light Microscopy, and Electron Microscopy). Make your subheadings general or more specific, depending on the type and amount of information you need to relate. If you use many specialized or invented terms to report your results, they may be put in the Methods section under Definitions. Using subheadings makes your text easier to write and to read, and it prods your memory for stray details on all aspects of the study.

■ Use specific, informative language.

Give your readers as much information as you can. Replace vague, imprecise words with more specific ones, and quantify your statements wherever possible.

VAGUE	I observed some monkeys in a large outdoor enclosure and others in small, individual indoor cages.
SPECIFIC	I observed 13 monkeys in an outdoor enclosure (10 $\times$ 8 $\times$ 12 m^3) and 12 others in individual indoor cages (1 $\times$ 2 $\times$ 1 m^3).
VAGUE	Several pits were dug at each forest site, and soil samples were collected from three different depths in each pit.
SPECIFIC	Four randomly located pits were dug at each forest site, and soil samples were collected from three depths at each pit: 0–5 cm, 6–11 cm, and 12–17 cm.
VAGUE	Root nodule tissue was stained with a number of histochemical reagents.
SPECIFIC	Root nodule tissue was stained with toluidine blue, Schiff's reagent, and aceto-orcein.
VAGUE	Every nest was checked frequently for signs of predation.
SPECIFIC	Every nest was checked twice daily (at 0800 and 1600) for signs of predation on eggs or nestlings.

■ Understand the difference between the active and the passive voice.

In the passive voice, the subject of the sentence *receives* the action:

Lizards (SUBJECT) *were collected* from three different sites.

However, in the active voice, the subject *performs* the action:

I (SUBJECT) *collected* lizards from three different sites.

Decisions about whether to use the passive or the active voice arise most often in the Materials and Methods section of research papers. Use of the passive voice is widespread in scientific writing, even though the active voice is generally more direct and concise. However, the passive voice focuses the reader's attention on the objects being studied or manipulated rather than on the researcher, and for this reason often seems more appropriate. Many writers find that a carefully constructed mix of active and passive sentences works best. See Chapter 7 (pp. 145–47) for more information on this subject.

■ Omit unnecessary information.

Include only those procedures directly pertaining to the results you plan to present. Do not get carried away in your desire to include all possible details. The reader is not interested in superfluous details or asides, and in published articles such material just wastes space and raises printing costs. The following example includes many unnecessary details:

> Fathead minnows were collected from Lost Lake, near Holmes, North Carolina, and transported back to the third-floor laboratory in large white pails. On the following Tuesday morning, skin for histological examination was taken from the dorsal part of the fish just behind the anterior dorsal fin.

Do we really need to know that the fish rode home from the lake in large white pails? Or that they were taken to that particular laboratory? Or that histological studies were done on a Tuesday?

Notice how much clearer the following example becomes when superfluous details are removed:

FAULTY	After considering a variety of techniques for determining the sugar content of nectar, I decided to use the method developed by Johnson (1955), because it seemed straightforward and easy to follow, especially for someone with a poor mathematical background.
REVISED	I used Johnson's (1955) method to measure the sugar content of nectar.

Do not confuse readers of your Materials and Methods section by insisting they jump ahead to your Results.

> To quantify the fright response, I observed 10 groups of fish, each composed of five individuals, and recorded the number of movements per 5-min interval (Fig. 1).

Remember that you are dealing strictly with procedures here. When readers encounter Figure 1 later in the paper, they will know enough to consult the Methods section if they need to.

RESULTS

■ Summarize and illustrate your findings.

The Results section should (1) *summarize* the data, emphasizing important patterns or trends, and (2) *illustrate and support* your generalizations with explanatory details, statistics, examples of representative (or atypical) cases, and tables and/or figures. To convey the results clearly, your writing must be well organized. Present the data in a logical order, if possible in the order in which you described your methods. Use the *past* tense (see pp. 147–48). If the Results section is long and includes many different topics, consider using subheadings to make the text easier for the reader to grasp.

The following paragraph is from a student paper on reproductive behavior in a species of damselfly. Notice that the author begins with a general statement and then supports it with quantitative data, a graph accompanying the text, and a selected example.

```
Observations of 21 marked males showed that the
number of matings per day varied among individu-
als. The number of matings per male ranged from
0 to 10 per day, with a mean of 6.5 ($s_{\bar{x}}$ = 3.2).
Males occupying territories with abundant emer-
gent vegetation encountered more females and
mated more frequently than males occupying
sparsely vegetated areas (Fig. 3). One male,
whose territory consisted entirely of open
water, obtained no matings during the five days
he was observed there.
```

■ Do not interpret the data or draw major conclusions.

The Results section should be a straightforward *report* of the data. Do not compare your findings with those of other researchers, and do not discuss why your results were or were not consistent with your predictions. Avoid speculating about the causes of particular findings or about their significance. Save such comments for the Discussion.

The following passage is from the Results section of a student paper on the algae present in a small stream. The writer begins with a statement of her findings (the first sentence), but then continues with interpretive material that really belongs in the Discussion.

```
The epilithic community was dominated by Achnan-
thes minutissima (Table 4). The abundance of
this species at the study site may be related to
its known tendency to occur in waters that are
alkaline (Patrick 1977) and well oxygenated (De-
Seve and Goldstein 1981). Its occurrence may
also have been related to . . .
```

Although generally biologists save interpretations of the data for the Discussion, you should note that some journals do allow a combined Results and Discussion section, followed by a short Conclusions section. In some studies, particularly those involving a series of successive experiments, readers can more easily understand the author's findings if one set of data is explained and interpreted before presentation of the next set. In such cases, the need for clarity should override a rigid adherence to conventional format. If an academic assignment seems to call for this sort of organization, check with your instructor.

▪ Integrate quantitative data with the text.

If the Results section includes tables or figures (discussed fully in Chapter 3), be sure to refer to *each* of these in the text. Do not excessively repeat in the text what is already shown in a table or figure, but also don't restrict yourself to passing comments ("Results are shown in Table 1"). Instead, point out the most important information or patterns and discuss them in the context of related data.

> Reproductive activity was closely related to time of day. Both males and females began to arrive at the breeding site in late morning, and the density of both sexes was highest between 1300 and 1500 h (Fig. 3). Data on air temperatures (Fig. 4) suggest that . . .

Do not automatically assume that anything involving numbers *must* be tabled or graphed in order to look important or "scientific." Unnecessary tables or figures take up space and waste the reader's time. Once you are sure which results are important and why, you may find that many can be summarized easily in a sentence or two. Following are some general guidelines about how to present numerical results verbally.

1. A mean value ($\bar{x}$) is often accompanied by the standard deviation, which gives the reader a sense of the variability of the data within the sample. Where consideration of the highest and lowest values is important, the range may be reported along with the sample size (number of observations). If you wish to show how reliable a sample mean is as an estimator of the population mean, give the standard error. See Chapter 2 (pp. 26–29) and Additional Readings (p. 193) for useful references on the use of these and other basic statistics.

The following sentences illustrate how statistical results can be integrated smoothly with the text:

> The 15 caterpillars in Group 3 averaged 2.1 cm in total length (range = 1.0).

> In the bullfrog choruses at Werner Lake, the mean size of central males was 140.42 mm ($s = 7.45$, $n = 12$).

You may also denote the standard deviation as follows:

> 140.42 ± 7.45 mm

2. When reporting the results of commonly used statistical analyses, you need not describe the tests in detail nor give all your calculations. Generally, you need to report only the major test statistics, along with the significance or probability level (see pp. 69–70).

> Analysis of variance showed significant variation among females with respect to mean egg mass ($F_{[29.174]} = 25.4$, $P < 0.001$).

3. As shown above, conventional abbreviations and symbols are used to report data succinctly within sentences. Some common ones are listed on the inside back cover of this book.

4. When do you write out the words for numbers, and when do you use numerals? The general rule is to use numerals when you (1) report statistics; (2) give quantitative data using standard units of measurement; and (3) refer to dates, times, pages, figures, and tables.

> The experiments were conducted between 1200 and 1500 on September 8, 9, and 10, 1987.

> Overlap of burrow groups ranged from 0 to 35%, and averaged 12% (Fig. 2 and Table 4).

In other situations involving numbers, use words for values of one through nine, and numerals for larger values. If you have a series that includes both values, use numerals for everything.

> Solitary individuals were two to three times more likely to encounter predators than were members of any of the five colonies.

> We followed the movements of 3 marked individuals, 22 marked pairs, and 14 unmarked pairs.

Do not *begin* a sentence with a numeral. Either write out the number or revise the sentence. If the number is part of a chemical term, it cannot be spelled out and the sentence should be reworded.

FAULTY	12 out of 425 eggs were cracked at the start of the first experiment.
REVISED	Twelve out of 425 eggs . . .
FAULTY	6-mercaptopurine was used to inhibit mitosis.
REWORDED	Mitosis was inhibited by 6-mercaptopurine.
REWORDED	I used 6-mercaptopurine to inhibit mitosis.

5. Watch your wording when you report quantitative results. Several commonly used words have restricted statistical meanings when they appear in scientific writing; do not use them loosely.

Significant: In popular usage, this word means "meaningful" or "important." Scientists use this term in a more restricted sense to refer to *statistically significant* (or nonsignificant) results, after having conducted appropriate statistical tests:

> Observed frequencies of turtles obtaining food differed *significantly* from expected frequencies ($\chi^2 = 58.19$, df = 8, $P < 0.001$).

Because the above results were significant, we know that the difference between observed and expected frequencies is probably a valid one. In this case, the likelihood of a difference of at least this magnitude occurring by chance alone is < 0.001, or less than one in a thousand. Note that often the leading zero is implied but not typed in probability values. Thus < 0.001 is often given as $< .001$. However, the Council of Biology Editors (CBE) prefers the inclusion of an initial zero preceding the decimal point for values smaller than 1.

When reporting the results of statistical tests, you always need to specify the significance level, as indicated by a probability value (*P*) such as the one above. Results associated with probabilities greater than 0.05 are generally considered *not* significant.

Correlated: In popular usage, two entities that are correlated are related to each other in some way. In scientific writing, the term *correlated* is used in conjunction with certain statistical tests (correlation analyses) that provide a measure of the strength of relationship between two numeric variables.

> Female size was not significantly *correlated* with the percentage of abnormally developing embryos in an egg mass ($r = 0.29$, $P < 0.05$, $n = 23$).

Note that correlation does *not* allow you to automatically assume a cause-and-effect relationship between the variables. It merely describes the extent to which they covary. For example, if variables X and Y are correlated, above average values of X tend to be associated either with above average values of Y (positive correlation) or with below average values of Y (negative correlation).

Random: In popular usage, this word means "haphazard" or "with no set pattern." However, scientists use this term to refer to a *particular* statistically defined pattern of heterogeneous values. If you write, "Subjects were assigned *randomly* to either Group A or Group B," this implies that you have used a table of random numbers or some other accepted method to make your group assignments truly random.

■ Omit peripheral information and unnecessary details.

Most scientists amass far more data than they ever present to their readers. Similarly, even in a weeklong project, you may have more results than you know what to do with. You may, understandably, feel reluctant to part with any of them. However, if you pack every last bit of information into the paper, you may lose sight of why you did the study in the first place. You must learn to present results selectively, to choose the data that are relevant to your hypothesis because they are either consistent or inconsistent with it. The data you present as results are the same data you will use to support your conclusions. You do not want to confuse the reader (or yourself) by including irrelevant information.

For example, the passage below is from a student field project on the diversity of vascular plants at a small pond.

    As shown in Figure 1, the shoreline of Hicks
    Pond was generally predominated by grasses and
    sedges. Cattails occupied a small area on the

```
northern end, and goldenrods (Solidago spp.)
were present in scattered groups along the east-
ern shore. A large population of red-spotted
newts (Motopthalmus viridescens) was also spot-
ted in this area, along with several species of
ducks. Species diversity was highest at Site D
(see Table 1), which was drier than most areas
and contained some of the same plants as in the
surrounding hayfield.
```

The third sentence in this paragraph should be omitted, because such zoological observations, however interesting, are irrelevant in this context.

Similarly, do not clutter the Results (or any other) section of the paper with irrelevant general statements of aims or intent. The following sentence, for example, should have been omitted from the student paper in which it appeared:

```
To present the results of this study, I will
first examine all of the relevant physiological
factors and then discuss the findings of each
feeding experiment.
```

You need not explain how you will proceed. If your writing is coherent and well organized, readers will follow your train of thought.

DISCUSSION

■ Interpret your results, supporting your conclusions with evidence.

In the Results section you reported your findings; now, in the Discussion, you need to tell the reader what you think your findings mean. Do the data support your original hypothesis? Why or why not? Refer to your data, citing tables or figures where necessary; use these materials as evidence to support your major argument or thesis. Here is the place, too, to discuss the work of other researchers. Are your findings consistent with theirs? How do your results fit into the bigger picture?

For many students, the Discussion section of a research paper is the most difficult portion to write. If you find yourself floundering, you may not fully understand what you are writing about yet. Take some time to rethink your study and why you did it. Look again at relevant literature on your topic. Then try drafting your ideas again. One of the chief benefits of writing and rewriting is to articulate your ideas ("writing to learn"). Be sure to leave ample time for this important process.

■ Do not present every conceivable explanation.

Sometimes beginners feel obliged to think of every possible way to interpret their results. The reader, swamped by explanations (most of them highly speculative), will quickly lose faith in the author. Look at this excerpt from a student Discussion section:

```
The dramatic decrease in Ochromonas in December
may have been related to the formation of cysts
(perhaps overlooked in the water samples taken
at that time). It is also possible that this
alga died off suddenly because of some environ-
mental stress, such as a prolonged period of
cold temperatures or a sudden chemical change in
the water. Light intensity and photoperiod are
other potential factors. Finally, the December
samples were not analyzed until the day after
they had been collected because two members of
our lab group were sick, so by this time the
algal populations may have been different from
the ones at the field site.
```

Remember that your task in the Discussion is to argue on behalf of the most plausible interpretations, based on the evidence available to you. To do so, you must be selective, focusing on explanations that have the greatest bearing on your study. Omit wild guesses and irrelevant asides.

■ Recognize the importance of "negative" results.

Experiments do not always have to confirm the presence of major differences, a strong relationship between two variables, or a conspicuous trend or pattern to be considered good science. Sometimes you may find no significant difference between two groups, or no relationship between a particular process and, say, temperature or some other factor you have been investigating. Such "negative" results still constitute respectable science, and they still need explanation. Thus, be alert to unexpected findings, and don't automatically conclude that the experiment is a failure or that you've made a mistake. Cases that do not conform to the expected pattern might represent something important — perhaps a new or altered focus for your study.

■ Make your prose convey confidence and authority.

Show that you are knowledgeable about your subject and take responsibility for your conclusions. Do not hedge or apologize. The tone in the following student examples is unnecessarily tentative:

```
This is just a preliminary study of the social
behavior of the guppy (Poecilia reticulata), but
it may possibly shed some light on the subject
and serve as a base for more conclusive work in
the future.
```

Would you want to read this paper? Perhaps not, if it is as preliminary and inconclusive as the author implies.

```
These results seem to suggest the possibility of
at least some connection between ferric chloride
and increased disease resistance.
```

This writer seems afraid to take a stand. What *do* the data show? If they suggest a connection between ferric chloride and disease resistance, then the writer should say so: "These results suggest a connection between ferric chloride and disease resistance."

```
These findings may not be very accurate because
of my limited experience, although they appear
to be consistent with the observations of Wilbur
(1972) and Morris (1978).
```

The last writer, like many other students, obviously feels intimidated by the work of "experts." However, "beginners" often do very good scientific work. Do not weaken your conclusions by unnecessary references to your status as a novice. If you are reluctant to make a strong statement, ask yourself why. Perhaps your anxiety stems from reservations about the data or the way the experiment was conducted. Perhaps there are certain variables you have not considered or certain assumptions that don't ring true. Those factors *are* worth considering, and it is important to sort them out from any general insecurities you have about yourself as a scientist or a writer.

■ Use a coherent, logical organization.

Instead of proceeding from the *general* to the *specific,* as the Introduction does, the Discussion moves from the *specific* to the *general.* There is no one right way to put together a Discussion, but the following plan is a common and effective one.

1. Start by drawing attention to your major findings without excessively repeating the Results. The reader has just looked at your data, so you do not have to describe them all over again. Beginning writers often start the Discussion by dredging up material from the Introduction; this approach, too, just adds redundancy to the text.

Focus the reader's attention on the most important findings, patterns, or trends. For example, here is the beginning of the Discussion section from a paper by Young, Greenwood, and Powell (1986, p 400):

> Although interactions between molluscs and their potential predators have been studied extensively (for reviews see Thompson 1960, Edmunds 1966, Todd 1981, Faulkner and Ghiselin 1983), these data are the first in which predator-prey interactions have been documented for a large number of potential predators of intertidal Onchidiids. In addition, our data indicate that the defensive secretion of *Onchidella borealis* has an effect on the distribution of the predatory seastar *Leptasterias hexactis*. The results indicate that *O. borealis* does not fire its repugnatorial glands in response to all potential predators, nor do all potential predators demonstrate flight behaviors in response to the glandular secretions of *O. borealis*.

2. Ask yourself what causes may underlie the major trends or phenomena you have described in the paper. If there are conflicting, problematic, or unexpected results, suggest explanations. Here is an excerpt from a student research report:

> The unusual presence of *Chaenorrhinum minus* in an untended garden (Site 5), instead of in its usual habitat of railroad cinder ballast, may be related to the use of railroad ties as decorative borders around the garden. Seeds from *C. minus* may have been lodged in crevices in the railroad ties, transported to the garden, and dislodged as the ties were being set in place. However, it is still unusual that *C. minus* was growing so well in a moist, shaded area among many other plants. My findings suggest that *C. minus* generally is a relatively poor competitor and grows best on dry, exposed, gravelly sites where few other plants are found.

3. Compare your findings with the work of other researchers. Are your results similar to theirs? At this point you can begin to supplement your own evidence with relevant findings from other studies, showing the reader how your work is part of a broader framework. Be sure to look at

the subject fairly and honestly. If some authors obtained results different from yours, point this out and suggest explanations for the differences. The following example is from McMillan and Smith (1974, p 52).

> The activities of males after spawning interested other observers of fathead minnows. Miller (1962) reported that a parental male positioned himself in the mouth of the cavity below the eggs and constantly fanned with his pectoral fins. This behavior has not been noted by other workers, nor has it been observed during the present study. It is possible that Miller was watching males hovering (our definition) below their eggs. In that case it is doubtful that the weak pectoral fin movements employed in hovering could substantially aerate the eggs — the function Miller attributed to "fanning."

4. End with more far-reaching predictions, interpretations, and conclusions. Can you generalize from your specific findings to other situations? How does your work contribute to an understanding of the broader topic? If you can end with a firm statement, as the student example below does, you give the reader a satisfying sense of closure.

```
In conclusion, cannibalism of eggs by larvae of
the butterfly Euphydryas phaeton occurred com-
monly under natural conditions, even within rel-
atively small colonies. These results are in
agreement with Wilson's (1975) suggestion that
detailed studies of cannibalism in animals may
show it to be more common than is usually sup-
posed. The finding that cannibalistic larvae
grew more rapidly than noncannibalistic ones
suggests that cannibalism may be an important
factor in larval development, especially when
food supplies are scarce. This nutritional bene-
fit of cannibalism may have long-reaching ef-
fects. If successful laboratory culture methods
can be developed for E. phaeton, we can further
explore the relationship between cannibalism and
individual fitness.
```

ACKNOWLEDGMENTS

A short Acknowledgments section usually comes between the Discussion and the References sections in a published research paper. In this section you thank anyone who assisted you with your research or with the preparation of your paper. For more on writing an Acknowledgments section, see page 160 in Chapter 8.

LITERATURE CITED

In the Literature Cited or References section, you list all the references *cited* (referred to) in your paper. Do not list any other sources, even if they proved to be useful background reading. See Chapter 6 for format guidelines.

SAMPLE RESEARCH PAPER

The following paper describes a student research project for an introductory biology course. References are listed using CBE format.

Germination of *Arabidopsis thaliana* Seeds
in Response to Exogenous Gibberellin
Treatments in Light and Darkness

Kristin A. VanderPloeg

Biology 121
April 21, 200-

Include title,
author, course,
and date on
title page for
student paper.
Do not number
title page, but
consider it page
1. When sub-
mitting a manu-
script for publi-
cation, follow
journal format
requirements.

Germination of *Arabidopsis thaliana* 2

ABSTRACT

Seed germination in plants is controlled by several mechanisms, including the presence of gibberellins as well as environmental conditions, such as temperature, light, and moisture. Germination of dormant seeds can be promoted by application of gibberellins, which may also act as natural endogenous regulators of the process. Here I report the effect of exogenous GA_3 on the germination of seeds of the *ga1-3* mutant of *Arabidopsis thaliana* in both light and darkness. Germination success increased significantly from 20.0% to 97.8% as the concentration of exogenous GA_3 increased from 0.001 to 0.5 mol L^{-1} in both light and dark conditions. However, there was no significant difference in germination success between the light and dark treatments. These data support the hypothesis that gibberellins are critical regulators of seed germination but are not consistent with previous studies showing stimulation of germination by light.

Key Words: gibberellin, gibberellic acid, germination, *Arabidopsis thaliana,* seed, dormancy, light sensitivity.

INTRODUCTION

Buried seeds of numerous plant species undergo a period of dormancy in response to environmental influences such as temperature, and lack of sufficient

---

Italicize genus and specific epithet. Capitalize genus only.

Summarize paper's most important contents using past tense. Use present to suggest a general conclusion.

Label and begin on same page as Abstract. In longer paper or journal manuscript, start new page.

Germination of *Arabidopsis thaliana* 3

moisture. In *Arabidopsis thaliana* (L.), dormancy allows seeds to survive the dry summer and germinate in autumn (Leon-Kloosterziel and others 1996). In an agricultural context, the establishment and subsequent breaking of dormancy may affect the efficiency of crop yield and the preservation of seeds for later planting. The actual mechanisms by which these adaptive responses are induced have been linked to various environmental and internal components such as light, temperature changes, and various members of a class of plant hormones called gibberellins (GAs); see Baskin and Baskin (1998) for a review.

Gibberellins exert profound effects on plant development. Numerous studies support the involvement of GAs in releasing seeds from dormancy (see, for example, Arnold and others 1996, Karssen and others 1989). Various forms of GAs have been identified and it is evident that the bio-active GA depends on the species of plant being examined (Derkx and others 1994). In *A. thaliana*, a pathway converts inactive GA_9 and GA_{20} to bio-active GA_4 and GA_1, respectively (Yamaguchi and others 1998). The nature of this conversion is the result of the expression of GA 3ß-hydroxylase (Yamaguchi and others 1998). GA_1 and GA_4 may facilitate germination by furthering the penetration of the seed coat by the radicle (Yamaguchi and others

Spell out *Arabidopsis* once, then abbreviate it when preceding species name, except at beginning of a sentence.

For background, briefly review pertinent literature. Name-year method is used here to cite sources. Use present tense to refer to established knowledge.

Germination of *Arabidopsis thaliana* 4

1998) or by producing an enzyme responsible for digesting tissue surrounding
the radicle tip (Groot and Karssen 1987).

The involvement of light in the
biosynthesis of GAs has also been the
topic of numerous studies. GAs can mimic
the effects caused by light in seed development and promote germination, and
exposure to light promotes the expression
of genes responsible for the synthesis of
GA 3ß-hydroxylase (Yamaguchi and others
1998). In the absence of this enzyme,
gibberellins GA_9 and GA_{20} are present but
are not converted to physiologically active forms. Indirectly, therefore, light
stimulates the synthesis of bio-active
GAs and thus promotes germination of
seeds of *A. thaliana.* This finding is
supported by work by Derkx and Karssen
(1993a), who found that light increases
the sensitivity of both wild-type and
ga1-2 mutant seeds of *A. thaliana* to a
mixture of GA_4 and GA_7. The *ga1-2* mutant
of *A. thaliana* cannot synthesize GAs and
therefore needs the application of exogenous GAs for germination. Derkx and
Karssen (1993a) also found that, in the
presence of light, the minimum concentration of GA needed to produce germination
is lower than the amount needed in darkness, showing an increase in seed sensitivity.

After giving rationale for study, state paper's objectives.

I examined the effect of the application of exogenous GA_3 in light and in

Germination of *Arabidopsis thaliana* 5

darkness to seeds of the *gal-3* mutant of
A. thaliana, which lacks the ability to
express GA_3-encoding genes and thus is dependent on exogenous GAs for germination.

MATERIALS AND METHODS

I obtained seeds of *A. thaliana*, mutant *gal-3*, from the Arabidopsis Biological Resource Center, Columbus, Ohio. Solutions of gibberellic acid (GA_3) were
prepared at concentrations of 0.001,
0.01, 0.1, and 0.5 mol L^{-1}. A 1-mL volume
of each GA_3 solution or water was added to
small petri dishes, each containing a
layer of filter paper and 10 *A. thaliana*
seeds. I prepared 19 sets of these 5
treatments. Of these, 10 sets were kept
under fluorescent "Gro-Lights" and 9 sets
were kept in a light-tight box. The number of germinated seeds was counted every
24 hours for 7 days. Radicle protrusion
was taken as the criterion for successful
germination. Water was added as needed
during the 7-day incubation period to
keep all seeds as close to the same moisture level as possible.

RESULTS

In the light, germination success of
gal-3 mutant seeds increased from 6% to
97.8% with increases in GA_3 concentration
from 0 to 0.5 mol L^{-1} (Table 1). A similar
trend with increasing GA_3 concentration
was observed in seeds maintained in darkness. A two-way analysis of variance
(Table 2) showed that the effect of GA_3

Use active voice when appropriate; use passive voice to focus attention on materials, not yourself.

Describe procedures in enough detail for others to repeat study.

Germination of *Arabidopsis thaliana* 6

Report statistical and other data. Use "*significant*" only for statistical significance.

concentration on germination success was highly significant ($P < 0.001$) but that there was no significant difference ($P > 0.05$) in germination success between seeds kept in the light and those kept in darkness. In addition, there was no significant interaction between GA_3 treatment and light conditions; thus, the significant effect of GA_3 treatment was the same in both light and darkness. Pairwise comparisons by means of Tukey tests showed that each successive increase in GA_3 concentration throughout the entire range produced a significantly greater germination success in both light- and dark-treated seeds.

Depending on instructor's preference, insert tables into text, close to where they are first mentioned, or else group them at end of paper.

Table 1. Germination success of seeds of the *gal-3* mutant of *A. thaliana* in response to treatment with GA_3 in light or darkness

GA_3 concentration (mol L^{-1})	Percent germination (mean $\pm$ 1 SE)	
	Light	Dark
0.001	22.2 $\pm$ 6.17	20.0 $\pm$ 9.28
0.01	40.5 $\pm$ 4.04	50.0 $\pm$ 6.68
0.1	87.7 $\pm$ 3.79	86.4 $\pm$ 3.37
0.5	97.6 $\pm$ 2.22	97.8 $\pm$ 2.29
0 (control)	6.0 $\pm$ 3.40	12.1 $\pm$ 5.73

Germination of *Arabidopsis thaliana* 7

Table 2. Two-way ANOVA on factors affecting germination success of *gal-3* mutant seeds of *A. thaliana*

Source of variation	df	MS	*F*
GA$_3$ concentration	4	19340.068	105.96***
Light condition	1	74.46	0.408^{ns}
GA$_3$ concentration × light condition	4	87.301	0.478^{ns}
Error	85	182.52	

*** = $P < 0.001$; *ns* = nonsignificant

DISCUSSION

Germination of *gal-3* mutant seeds of *A. thaliana* increased with increasing concentrations of exogenous gibberellin. Similar findings have been reported for other *A. thaliana* lines (Derkx and Karssen 1993a, 1993b; Derkx and others 1994) and with dormant seeds of *Chaenorrhinum minus* (Arnold and others 1996). All of these authors concluded that endogenous GAs may be responsible for breaking seed dormancy and stimulating germination in wild-type seeds under natural conditions.

My results show that seeds germi-

Briefly review pertinent findings. Do not excessively repeat results.

Relate findings to others'. Shift to present tense to discuss previously published information.

Germination of *Arabidopsis thaliana* 8

nated equally well in light and in darkness. However, similar studies (Derkx and
Karssen 1993a; Derkx and others 1994)
showed that light increases endogenous GA
levels and, in turn, increases germination success of seeds of the *gal-2* mutant
of *A. thaliana.* In contrast to my results, however, these authors found that
seeds incubated in the presence of exogenous GAs and kept in darkness do not germinate. Although the precise effects of
light depend on the temperature and on
the season in which seeds are harvested,
Derkx and Karssen (1993b) found that an
exposure to red light for as little as 15
min, followed by incubation in darkness,
is sufficient to cause an increase in
germination success. Peng and Harberd
(1997) suggest that this sensitivity is
due to the presence of phytochromes,
which are light-sensitive plant pigment
molecules that regulate many light-
dependent processes in plants including
seed dormancy and germination. The active
form of a specific phytochrome (PHYB)
promotes low frequency seed germination
after an initial exposure to red light
(Yamaguchi and others 1998). Further work
is necessary to determine reasons for the
discrepancy between my results and those
of Derkx and Karssen (1993a, 1993b) and
Derkx and others (1994). A possible explanation lies in the different mutant
strains used. It is also possible that

Germination of *Arabidopsis thaliana* 9

germination in the dark treatments was a response to the brief exposure to fluorescent light during application of control and gibberellin solutions and before seeds were placed in the light-tight box. Since this exposure was of short duration (less than 30 min) and given in the very early stages of imbibition, this explanation may be unlikely, although such responses are known for seeds with extreme light sensitivity (Kendrick and Cone 1985).

ACKNOWLEDGMENTS

I thank C. Visscher for advice during the research and preparation of this paper.

LITERATURE CITED

Arnold RM, Slyker JA, Gupta TH. 1996. Germination of *Chaenorrhinum minus* seeds in response to gibberellin treatments. J Plant Physiol 148:677-83.

Baskin CC, Baskin JM. 1998. Seeds: ecology, biogeography, and evolution of dormancy and germination. New York: Academic Press. 666 p.

Derkx MPM, Karssen CM. 1993a. Effects of light and temperature on seed dormancy and gibberellin-stimulated germination in *Arabidopsis thaliana*: studies with gibberellin-deficient and -insensitive mutants. Physiol Plant 89:360-8.

Derkx MPM, Karssen CM. 1993b. Variability

With their permission, acknowledge those who assisted with project.

Include all references cited. Alphabetize sources by first author's last name. Here, follow CBE format.

For a book, give author(s), publication date, title, place of publication, publisher, and number of pages.

Use initials for authors' first and middle names followed by publication date, title of paper with only first word capitalized. Use conventional abbreviations for journal title. Give volume number followed by page numbers.

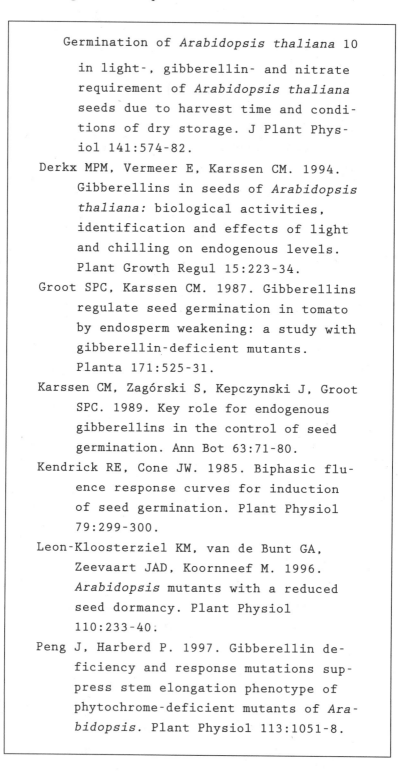

Germination of *Arabidopsis thaliana* 10
in light-, gibberellin- and nitrate
requirement of *Arabidopsis thaliana*
seeds due to harvest time and conditions of dry storage. J Plant Physiol 141:574-82.

Derkx MPM, Vermeer E, Karssen CM. 1994.
Gibberellins in seeds of *Arabidopsis
thaliana:* biological activities,
identification and effects of light
and chilling on endogenous levels.
Plant Growth Regul 15:223-34.

Groot SPC, Karssen CM. 1987. Gibberellins
regulate seed germination in tomato
by endosperm weakening: a study with
gibberellin-deficient mutants.
Planta 171:525-31.

Karssen CM, Zagórski S, Kepczynski J, Groot
SPC. 1989. Key role for endogenous
gibberellins in the control of seed
germination. Ann Bot 63:71-80.

Kendrick RE, Cone JW. 1985. Biphasic flu-
ence response curves for induction
of seed germination. Plant Physiol
79:299-300.

Leon-Kloosterziel KM, van de Bunt GA,
Zeevaart JAD, Koornneef M. 1996.
Arabidopsis mutants with a reduced
seed dormancy. Plant Physiol
110:233-40.

Peng J, Harberd P. 1997. Gibberellin de-
ficiency and response mutations sup-
press stem elongation phenotype of
phytochrome-deficient mutants of *Arabidopsis.* Plant Physiol 113:1051-8.

Germination of *Arabidopsis thaliana* 11

Yamaguchi S, Smith MW, Brown RGS, Kamiya Y, Sun T. 1998. Phytochrome regulation and differential expression of gibberellin 3ß-hydroxylase genes in germinating *Arabidopsis* seeds. Plant Cell 10:2115-26.

CHAPTER 5

Writing a Review Paper

A scientific review paper is a critical synthesis of the research on a particular topic. Biologists read review papers to stay abreast of the current knowledge in a field and to learn about subjects that are unfamiliar to them.

In writing review papers, authors vary in their purposes and scope. For example, we might classify biological review papers into at least five different types:

A *state-of-the-art review* presents an up-to-date, interpretive synthesis of our knowledge of a certain subject or issue, with emphasis on the most recent literature. For example, an author might look at what is currently known about the advantages and disadvantages of a particular surgical method for mastectomy, or a particular chemotherapy regime in the treatment of breast cancer.

A *historical review* examines the historical development of knowledge about a particular topic. It may focus on the emergence of the subject area as an important field of study, the major pioneering work by early contributors, the replacement of outdated theories or perspectives, and the development of improved methodologies. For example, an author might survey our growing understanding of territoriality in birds, beginning with the earliest ideas and (perhaps) misconceptions, and ending with the most recent models and problems.

A *comparison-of-perspectives review* critically examines two or more ways of looking at a particular scientific issue. For example, an author might ⌐studies supporting the idea that the transport of sugars in phloem ⌐ve metabolic process with studies supporting the hypothesis ⌐cess by mass flow.

⌐*-fields review* attempts to provide new insights about y bringing in literature from another area or discipline.

For example, a reviewer might examine how the language and methology of a particular branch of microeconomics has shed light on our knowledge of reproductive strategies in insects.

A *theoretical or model-building review* surveys the literature on a particular scientific problem — for example, carbon allocation in trees, or cooperative breeding in birds — with the aim of presenting a new theoretical context or an alternative model.

These arbitrary categories are not rigid, of course, and many reviewers ambitiously and successfully fuse more than one approach in their investigation and subsequent expression of a particular topic. Because of possible differences in emphasis, the format of a review is not as standardized as that of a research paper. Most reviews have an Introduction and a Discussion or Conclusions section, and many also have an Abstract. Long reviews may be preceded by a Table of Contents. All end with a lengthy Literature Cited section.

When you are assigned to write a review paper for a biology course, you need to rely on many of the same strategies used by authors of published reviews. You are expected to address readers similar to yourself and your classmates — people with a background in the broad subject area but without specialized knowledge of your particular topic. Your aim is not only to inform but also to evaluate and interpret. A good review bears the stamp of the writer's own thought processes.

CHOOSING A TOPIC

Do not underestimate the importance of choosing a suitable topic. Ideally, it should be (1) interesting to you; (2) not so broad that it is unmanageable; (3) not so narrow you can't find enough information on it; (4) not so difficult that you can't fully understand it.

A common approach is to start with a broad topic and do some general reading about it, gradually narrowing it down to a workable size. Suppose you decide to write a paper about orchids. Because orchids are a large, diverse group and much has been written about them, you will need to restrict yourself to some specific aspect of these plants. For example, you might decide to write about the unusual interactions between orchids and their insect pollinators. Eventually you might narrow your topic even further, perhaps confining yourself to only a few species or to several, similar kinds of pollination interactions.

As you narrow your topic and become familiar with the literature, you need to develop a sense of your main objectives. What question will your paper address and from what perspective? Are you shifting toward a particular viewpoint or conclusion that can serve as a main point, or thesis, for the paper? If you limit your scope and define your goals early in the project, your reading and note taking will be more directed and your time will be more productive.

Occasionally the first topic you choose may be *too* narrow and you will have to enlarge it or shift to a different subject altogether. Allow time for false starts, delays, and topic changes. Recognizing a good subject to write about requires thought and effort.

DECIDING ON A TITLE

Like a research paper, a review paper should have a specific and informative title. If your title is vague or too general, it will suggest that you lack a firm sense of your paper's aims and scope. If you are preparing a paper for publication, remember that many indexing and abstracting services rely on key words in titles in their organization of the journal literature. Your title should therefore also be concise; cut any words that do not serve an important function in describing the paper's contents. Consider the following sets of examples:

VAGUE	Studies on Seed Coats
IMPROVED	The Role of Seed Coats in Seed Viability
VAGUE	The Bromeliaceae: Research Questions and Controversies
IMPROVED	Physiological Ecology of the Bromeliaceae
WORDY	A Review of the Literature on the Functions of the Avian Spleen
IMPROVED	Functions of the Avian Spleen: A Review
WORDY	Strategies of Seed Dispersal by Plants Inhabiting Desert Environments
IMPROVED	Seed Dispersal Strategies in Desert Plants

See also pages 52–55 for more on titles.

WORKING WITH THE LITERATURE

A review paper involves a thorough search of the literature on your topic, followed by careful reading and selective note taking. These parts of the research process are covered in Chapter 1. However, it bears repeating here that constructing a good review requires the ability to be discriminating in your use of materials. Remember that a review paper is a secondary source that provides a critical, focused synthesis of *primary* sources. In biology, the primary literature consists largely of research papers in scientific journals, along with government or technical reports, conference proceedings, and other works presenting original research to a scientific audience. It is perfectly acceptable to *start* your research with scientific encyclopedias, textbooks, lab manuals, nontechnical books, and articles in

Discover, National Wildlife, Scientific American, Natural History, and similar magazines for lay readers. All these sources provide excellent background reading and help prepare you for the more specialized papers you will need to read later. Scientific review papers in your general subject area also serve this function. However, secondary references should not be major sources of information for your paper.

For example, suppose you are writing a review paper on the visual system of horseshoe crabs — a topic you chose after coming across an article titled "What the Brain Tells the Eye" by Robert B. Barlow, Jr., in *Scientific American*. The article contains much information that would be useful for your paper, and like other features in the magazine, it is written by an expert on the topic. However, you need to go one step further. Rather than using the *Scientific American* article as a major source for your paper, you should track down some of Barlow's original research findings in the scientific papers in which they first appeared. For example, in the list of Further Reading following the *Scientific American* article is a paper by Barlow and three coauthors titled "Circadian Rhythms in *Limulus* Photoreceptors: I. Intracellular studies," in *Journal of General Physiology*. With the background knowledge you've acquired by reading Barlow's more general article, you should feel ready to tackle this primary source, along with other specialized research papers. Such references make up the primary literature on vision in horseshoe crabs.

In summary, for a biological review paper you need to focus on papers in scientific journals, along with other primary sources. Rely on secondhand information as little as possible; cite these materials rarely, if at all. If you are unsure about how to distinguish primary from secondary sources, or about how to locate primary sources, review Chapter 1, or seek help from your instructor or your reference librarian. With time, you'll become familiar with the important journals featuring research on your topic, and probably, too, with the names of key investigators in the field. Remember that the quality of your review paper will depend largely on the sophistication of your literature search, your ability to make informed choices among a variety of possible references, and your mastery of key primary literature.

PRESENTING YOUR MATERIAL

■ Sketch out a rough plan or make an outline.

Biological review papers are not as standardized in their format as research papers; their organization depends on the subject, the writer's objectives, and (in the case of published articles) the editorial guidelines of the journal. Most reviews have an Introduction, a body (not labeled as such, but often with headings and subheadings), a Conclusions section, and a Literature Cited section. Some have a Summary at the end. Many start with a short Table of Contents that functions as an outline to the paper. Some have an Abstract, which, like the Summary, states the purpose and scope of the review and summarizes the major points covered. As in a research paper, the Abstract is placed after the title and author, before the main text of the paper. (See pp. 55–59 for more on Abstracts.)

For academic assignments, your instructor may have preferences about the format and components of the paper. If not, a good strategy is to model your own paper after a published review in your general subject area.

Many journals, such as *Biological Bulletin* and *American Midland Naturalist,* publish review articles in addition to research papers. Other journals specialize entirely in reviews; these include *Botanical Review, Physiological Reviews, Psychological Review, Quarterly Review of Biology,* and various annually published volumes of reviews, such as *Annual Review of Ecology and Systematics, Annual Review of Microbiology,* and *Annual Review of Genetics.*

Before you plunge into the first draft, you will need a tentative plan. This can range all the way from a rough sketch of the order of topics to a formal, detailed outline. Some people feel hemmed in by outlines and prefer to do much of their organizing as they work out the first draft. (These are the writers who, if ever asked to produce an outline, do so *after* they have written at least one draft of the paper.) Other people depend on a fixed plan right from the beginning to organize their thoughts. There is no one correct way to plan a review paper. You must decide what kind of organizational method suits your own writing style.

At some point in the writing, you may wish to use your outline or plan to sketch out a Table of Contents, which in long review papers is an enormous aid to the reader. (Short reviews can do without a Table of Contents as long as the material is still well organized.) Here is the Table of Contents for a student review paper in animal behavior:

THE ADAPTIVE SIGNIFICANCE OF ALARM CALLS IN MAMMALS

Note the use of headings and subheadings in this paper to give order to an otherwise unwieldy amount of material. Subdividing the text may also make the paper easier to write because you can tackle one chunk of material at a time. See pages 158–59 for how to compose and format headings in your text.

■ Introduce the subject, explain your rationale, and state your central question, objectives, or thesis.

These three tasks need to be accomplished in the Introduction. As in a research paper, an effective strategy is to start with broad statements, explanations, and definitions that orient and educate the reader. Then work down to more specific issues. Why is this subject important? What approach have you taken? Will you be giving a comprehensive summary or one that is more limited? End the Introduction with a clear statement of the question you will address or the main point you wish to convey to the reader. Your objectives may be fairly specific — for example, to show that certain kinds of early childhood experiences predispose adolescent girls toward anorexia. Or you may wish to assess the current state of research on a particular problem, for instance, current treatment methods for AIDS, with the aim of making predictions about the next decade.

The length of the Introduction depends on your subject and the kind of coverage you plan to give it. An 8- to 10-page review might need only a single succinct paragraph to introduce it. A longer paper might require a whole page. Generally, one or two paragraphs are appropriate. Readers will soon become confused if you do not tell them your objectives fairly early in the paper.

Here is the Introduction from the first draft of a term paper for a molecular biology course. The numbers in parentheses refer to sources in the Literature Cited section (see discussion in Chapter 6). How might this student paragraph be improved?

```
Many different articles were read about the mo-
lecular genetics of human growth hormone. This
paper will focus especially on hGH deficiencies.
Human growth hormone (hGH) is a polypeptide hor-
mone, produced from within a gene cluster on
chromosome 17, that controls much of the physi-
cal growth of the infant and child (1, 2). Since
time is limited, this paper cannot cover all
possible aspects of hGH, so a narrower approach
has been taken.
```

In this Introduction, two sentences, the first and the last, say nothing essential. The reader assumes you have read about your topic, and you can easily show how you plan to narrow the focus by stating exactly what this focus is. Remember that the Introduction gives you your first chance (perhaps your only chance) to interest the reader. Obviously your instructor must read the paper whether he or she wants to or not, but if your start is forceful and interesting, the paper will have a much better effect.

A second problem with the paragraph is that we don't get a clear sense of the writer's *specific* purpose or rationale. We know that the paper will focus mainly on hGH deficiencies, but we do not know how or why. If the writer fails to portray the subject as important or intriguing, it is difficult for the reader to feel it is.

Here is the second-draft version of the Introduction. Notice how the writer has omitted the unnecessary sentences and filled in the gaps by expanding the more important parts of the original. The revised Introduction conveys a clearer idea of what this paper is about and why this subject is interesting.

```
Human growth hormone (hGH) is a polypeptide hor-
mone, produced from within a gene cluster on
chromosome 17, that controls much of the physi-
cal growth of the infant and child (1, 2). Defi-
ciency of hGH, a heritable disorder, can result
in infantile dwarfism and retardation (3, 4, 5).
New research methods, including recombinant DNA
technology, have made it possible to determine
the molecular basis of such deficiencies. In
this paper, I will summarize current knowledge
of the molecular genetics of hGH and suggest
ways in which continued research may help physi-
cians treat infants with a deficiency of this
hormone.
```

■ Build a focused discussion.

Many student review papers are little more than summaries, boring ones in which the writer has retreated from the reader's sight. Just to regurgitate the contents of a series of papers one by one is not enough. You need to *relate* this material to your principal objectives. Present your information selectively, and use it to support or illustrate the statements you wish to make. A good review interprets the literature from the writer's own informed perspective and gives the reader a sense of integration, development, and focus.

The following paragraph is from a paper on the adaptive value of cannibalism in animals. Notice how the author uses examples from the literature to illustrate and develop the generalization in the first sentence.

```
Following the reasoning of West Eberhard
(1975), we may predict that cannibalism may
be more likely when the potential victims are
```

unusually vulnerable and easily obtained as
food. Such individuals are, in fact, the
predominant victims of cannibalism in many
species. For example, most cannibalism in
Tribolium is performed by larvae and adults
on the defenseless eggs and pupae (Mertz and
Davies 1968). Pupae are also eaten by larvae
in caddisflies (Gallepp 1974), and injured or
weak immature stages are devoured by older
nestmates in many species of social ants
(Wilson 1971). In crows (*Corvus corone*), cannibalism of eggs and nestlings by intruding
adults is more frequent when the parents are
absent from the nest, leaving their young
more vulnerable to attack (Yom-Tov 1974).

The next passage is from a paper for a plant pathology course. It discusses host-specific toxins, substances produced by pathogenic (disease-
producing) fungi that attack certain plants. Notice that the author presents
selected information from the literature to critically examine a particular
hypothesis. The writing conveys authority and a thorough familiarity with
the material.

Changes in the permeability of host cell
membranes after being exposed to toxins suggest
that these substances may bind to a receptor
site in the cell membranes of susceptible hosts.
Strobel (1974) claims to have found such a receptor site in the membranes of sugarcane cells
treated with helminthosporoside, a toxin isolated from the pathogenic fungus *Helmintho-
sporium sacchari.* The site contains a protein of
molecular weight 48,000 daltons that specifically binds the toxin. Strobel has proposed that
the binding of the toxin stimulates the activity
of an adjacent membrane-bound enzyme, potassium-
magnesium ATPase, which maintains ion balance
across cell membranes. Such stimulation could
disrupt the membrane's selectivity to ions,

resulting in the characteristic symptom of elec-
trolyte leakage.

Several authors have criticized Strobel's
methodology and interpretations. For example,
Wheeler (1976) doubts that the preparation used
for helminthosporoside assay and structure determination was sufficiently free from impuri-
ties, and he argues that some experiments lacked
sufficient replication. Others have said that
the toxin-binding data were not graphed correctly and that regraphing them suggests the
binding activity is, at best, weak (see, for example, Hanchey and Wheeler 1979). These and
other criticisms have been reviewed by Yoder
(1980).

■ Document your paper thoroughly.

Whenever you refer to another author's work or ideas, cite
your sources using conventional methods of literature citation (see
Chapter 6). In a review paper (as in the Introduction and Discussion
of a research paper), you need to cite references repeatedly. Look at
the following sentence from a review by Hepler and Wayne (1985,
p 412).

> Red light triggers a large array of physiological and developmental
> events that require Ca^{2+}, including chloroplast rotation in
> *Mougeotia* (55, 78, 247–249), spore germination (254–256) and
> cell expansion (37) in *Onoclea*, leaflet closure in *Mimosa* (22, 23,
> 237), root tip adhesion in *Phaseolus* (229, 279), peroxidase secre-
> tion in *Spinacia* (113, 164, 165), membrane depolarization in
> *Nitella* (261), as well as activation of NAD kinase (1, 218, 232)
> and inhibition of mitochondrial ATPase (212).

Using many citations may seem strange at first. You may feel they impede the flow of your writing. However, readers of scientific papers are
accustomed to such interruptions, and you will get used to them, too. Remember that literature citations serve an important function: they tell
readers where to find additional information. Careful documentation also
reflects the thoroughness of your literature search and your honesty in acknowledging the sources of your material.

■ Use quoted material sparingly.

Many beginning writers, unsure of their own voices or uncomfortable with the material, tend to fall back on quotations to get them through rough spots in the paper. Sometimes they construct whole paragraphs around a series of quotations from different authors, stringing these together with a few scattered phrases or sentences of their own. The text thus becomes a collection of other people's words:

> Studies of the Baltimore butterfly (*Euphydryas phaeton*) showed that "larvae occupied communal nests of various sizes" and "commonly cannibalized unhatched eggs in the same colony" (Monti 1950). "In both field and laboratory tests, there was a higher incidence of cannibalism by larvae occupying large colonies" (Monti 1950). Also, "cannibalistic acts occurred at a higher frequency under conditions of food shortage," when the larval foodplant, turtlehead, "was in short supply or extensively defoliated" (Mulry 1960).

In such a passage the reader loses track of the writer and the writer loses authority. The quotations do not enhance the text — they detract from it, suggesting that the writer hasn't come to terms with the material and is either too inexperienced or too lazy to use his or her own words. Biological authors rarely use quoted material, relying instead on careful, concise paraphrases or summaries; you should do the same. For example, the quotation-ridden passage above can be rephrased and condensed in the author's own words:

> Studies of the Baltimore butterfly (*Euphydryas phaeton*) showed that cannibalism of eggs by the communal larvae was more frequent in large colonies than in small ones (Monti 1950). Cannibalism also increased when the larval foodplant, turtlehead, was scarce (Mulry 1960).

When *is* it appropriate to use the exact wording of an author? Occasionally you may wish to include a quotation to establish or emphasize an important point or state a precise definition. Sometimes you may feel that an author's exact words are indispensable in conveying a particular viewpoint or idea:

> Bem (1981, p 255) defines a *schema* as a "cognitive structure, a network of associations that organizes and guides an individual's perception."

> To Eiseley (1961), Darwin was "the man who saw the wrinkled hide of a disintegrating planet, glyptodonts and men, all equally flowing down the direction of time's arrow; he was a master artist and he entered sympathetically into life" (p 351).

In his controversial book *The Naked Ape,* Morris (1967, p 211) concludes that the survival of the human species depends on an increasing awareness of our biological heritage:

> We must somehow improve in quality rather than in sheer quantity. If we do this, we can continue to progress technologically in a dramatic and exciting way without denying our evolutionary inheritance. If we do not, then our suppressed biological urges will build up and up until the dam bursts and the whole of our elaborate existence is swept away in the flood.

When you do use quoted material, cite the source in the text using either the name-year or number method (see Chapter 6). Some biological authors include the page number of the book or article in the citation; others do not. Be sure to reproduce the quoted material *exactly*. Introduce and punctuate quotations properly, using the following rules.

1. Do not substitute single quotation marks (' ') for double ones (" "). The former are restricted to quotations within other quotations. (In British usage, single quotation marks are used first; double quotation marks are used to set off quotations within quotations.)

2. A short quotation can be integrated into the text. Make sure it fits in grammatically with the rest of the sentence.

> Dawkins (1976, p 206) suggests that memes, as units of cultural transmission, can replicate themselves by "leaping from brain to brain via a process which, in the broad sense, can be called imitation."

For a quotation longer than four typed lines in your paper, omit the quotation marks and indent each line of the passage five spaces from the left margin, but keep it double-spaced. A colon is often used to introduce the passage.

3. Place periods and commas *inside* the quotation marks; semicolons and colons go *outside*.

> Jones (1987, p 2) calls Davidson's explanation "the most exciting model of this century."

> Davidson's explanation is "the most exciting model of this century," according to Jones (1987, p 2).

> Jones (1987, p 2) considers Davidson's explanation "the most exciting model of this century"; unlike previous models, it gives rise to far-ranging predictions about how major evolutionary changes occur.

If exclamation marks, question marks, and dashes are part of the quoted material, put them inside the quotation marks. If they are part of your own sentence, put them outside.

4. If you need to interrupt a quotation to omit one or more words, indicate the omission by an ellipsis (three periods, separated by spaces). If

the ellipsis falls at the end of a sentence, put the period followed by a space before the ellipsis.

> Morris (1976, p 211) concludes that humans must "somehow improve in quality. . . . If we do this, we can continue to progress technologically in a dramatic and exciting way without denying our evolutionary inheritance."

5. If you insert clarifying or explanatory material into a quotation, put such material in brackets [ ].

> Morris (1976, p 211) concludes: "If we do this [improve in quality], we can continue to progress technologically in a dramatic and exciting way without denying our evolutionary heritage."

■ End with general conclusions.

Despite the importance of an effective Conclusions section, it gets the least attention from many beginning writers of a review paper. By the time they reach this point, most writers are worn out and feel they have nothing more to say. They are so anxious to be done with the paper that they make a hasty retreat, summing up the topic in a sentence or two and then coming to a weary stop.

> Thus the role of host-specific toxins in plant disease is a complex topic. There are still many questions for scientists in the future to answer.

The following student paragraph, however, summarizes the major points in the paper and adds a final perspective:

```
In conclusion, current evidence shows that re-
sistance or susceptibility is an inborn trait of
a plant. Some plant pathogenic fungi can produce
toxins that, by disturbing host physiological
functions, affect the onset and development of
plant disease. However, it is not clear whether
these substances have a primary role in deter-
mining resistance or susceptibility. Perhaps
some other mechanism may determine specificity,
and toxins may exert their damaging effects sec-
ondarily. Future research on host-specific tox-
ins will help to answer this question.
```

As shown above, a Conclusions section consolidates and strengthens the relationships, patterns, and arguments you have been building in the body of the paper. In addition, a good Conclusion should do more than merely summarize; it must also *conclude* something. Assuming that you have adequately addressed the topic, now you must answer such questions as "So what?" or "What next?" What is the significance of everything you have just told the reader? What conflicts still need to be resolved? What research must still be done? What might we expect to happen in the future? Notice how the student writer of the following passage has begun to consider such issues in her review of intervertebral disk herniation (IVD) in the dog:

> Within the past decade, some of the controversy regarding the use of fenestration, dorsal laminectomy, and hemilaminectomy for treating IVD herniations has been resolved. Evidence now suggests that routine fenestration of the affected disc, coupled with the selective use of prophylactic fenestration, may be the safest and most beneficial means of lowering recurrence rates for IVD disease. However, far more research is needed to identify and characterize the risk factors that predispose a patient to multiple, and in particular adjacent, disc extrusions. Accordingly, study designs will need to become more consistent both in methodology and in definition of terms....

Do not, however, introduce *new* information that really belongs in the body of the paper. Avoid complex questions or issues that you can't fully address. You do not want to throw readers off course or leave them hanging in midair. On the contrary, you want to tie up loose ends and finish with a satisfying sense of closure.

There is no set length for a Conclusions section. In published reviews, it varies from a single concise paragraph to a page or more of text. The length depends on the topic, the author's aims and depth of coverage, and any length restrictions on the paper. Plan ahead as you are drafting the manuscript so that you have enough space, time, and energy to end your paper effectively.

SAMPLE REVIEW PAPER

The following excerpt is from a student review paper for a physiology course. Literature Cited or References are listed using the CBE number system style of documentation (see p. 197).

The Role of Hypothermia
and the Diving Reflex in Survival
of Near-Drowning Accidents

Brian Martin

Biology 281
April 17, 200-

Include title,
author, course,
and date on
title page for
student papers.
Do not number
title page, but
consider it
page 1.

Hypothermia and Diving Reflex 2

State aims and scope; concisely summarize major points.

ABSTRACT

This paper reviews the contributions of hypothermia and the mammalian diving reflex (MDR) to human survival of cold-water immersion incidents. The effect of the victim's age on these processes is also examined. A major protective role of hypothermia comes from a reduced metabolic rate and thus lowered oxygen consumption by body tissues. Although hypothermia may produce fatal cardiac arrhythmias such as ventricular fibrillation, it is also associated with bradycardia and peripheral vasoconstriction, both of which enhance oxygen supply to the heart and brain. The MDR also results in bradycardia and reduced peripheral blood flow, as well as laryngospasm, which protects victims against rapid inhalation of water. Studies of drowning and near-drowning accidents involving children and adults suggest that victim survival depends on the presence of both hypothermia and the MDR, as neither alone can provide adequate cerebral protection during long periods of hypoxia. Future lines of research are suggested and related to improved patient care.

Introduce topic; give paper's aims and scope. CBE number system is used here to cite sources, which are numbered in order of mention in the text.

INTRODUCTION

Drowning and near-drowning incidents are leading causes of mortality and morbidity in both children[1] and adults.[2] Over the past 30 years, there has been

Hypothermia and Diving Reflex 3

considerable interest in cold-water immersion incidents, particularly the reasons for the survival of some victims under seemingly fatal conditions. Research suggests that both hypothermia and a "mammalian diving reflex" (MDR) may account for survival in many near-drowning episodes.[3] However, the extent to which these two processes interact is not fully understood. Controversy also exists regarding the effect of the victim's age on the physiological responses to cold-water immersion. In this paper, I provide an overview of recent research on the protective value of hypothermia and the MDR in cold-water immersions. I also examine hypotheses concerning the effects of age on these processes, and conclude with suggestions about future lines of research that may lead to improved patient care.

Hypoxia during drowning and near-drowning incidents

The major physiological problem facing drowning victims is hypoxia, or lack of adequate oxygen perfusion to body cells.[1,4] Hypoxia results in damage to many organs, including the heart, lungs, kidneys, liver, and intestines.[4] Generally, the length of time the body has been deprived of oxygen is closely related to patient prognosis. Only 6-7 s of hypoxia may cause unconsciousness; if

In body of paper, summarize and integrate data from primary sources. Use headings, if desired, to organize text.

Hypothermia and Diving Reflex 4

hypoxia lasts longer than 5 min at relatively warm temperatures, death or irre-
versible brain damage may result.[5] However, some victims of immersion in cold
water have survived after periods of
oxygen deprivation lasting up to two
hours.[4] . . .

[The student goes on to highlight the major controversies and to add interpretation and analysis.]

Use original
number as-
signed to refer-
ence when cit-
ing it again.

CONCLUSIONS

Conclude with
suggestions for
future research,
links to broader
issues.

Recent research on cold-water immersion incidents has provided a more complete understanding of the physiological
processes occurring during drowning and
near-drowning accidents. Current findings
suggest that the cooperative effect of
the MDR and hypothermia plays a critical
role in patient survival during a cold-
water immersion incident. However, the
relationship between the two processes
is still unclear. Because it is impossible to provide an exact reproduction of
a particular drowning incident within
the laboratory, research is hampered
by the lack of complete details surrounding drowning incidents. Consequently,
it is difficult for comparisons to
be drawn between published case
studies.

More complete and accurate documentation of cold-water immersion incidents —
including time of submersion; time of
recovery; and a profile of the victim

Hypothermia and Diving Reflex 5

including age, sex, physical condition —
will facilitate easier comparison of individual situations and lead to a more
complete knowledge of the processes affecting long-term survival rates for
drowning victims. Once we have a clearer
understanding of the relationship between
hypothermia and the MDR — and of the effect
of such factors as the age of the victim —
physicians and rescue personnel can take
steps to improve patient care both at the
scene and in the hospital.

ACKNOWLEDGMENTS

I would like to thank V. McMillan
and D. Huerta for their support, suggestions, and patience throughout the production of this paper. I am also grateful
to my classmates in COMP 281, for their
thoughtful comments during writing work-
shops. Finally, I thank Colgate's interlibrary loan staff for their tireless
efforts to secure the sources I needed
for this paper.

With their permission, acknowledge people who assisted with the paper.

LITERATURE CITED

Include all references cited. Numbers correspond to order in which references were first mentioned in the text. For journal articles, CBE format calls for author's last name, followed by initials; then article title, abbreviated journal title, publication date, volume or issue number, and inclusive page numbers.

1. Kallas HJ, O'Rourke PP. Drowning and
   immersion injuries in children. Curr
   Opin Pediatr 1993;5:295-302.

2. Keatinge WR. Accidental immersion hypothermia and drowning. Practitioner
   1977;219:183-7.

3. Gooden BA. Why some people do not
   drown — hypothermia versus the diving
   response. Med J Aust 1992;157:
   629-32.

Hypothermia and Diving Reflex 6

4. Biggart MJ, Bohn DJ. Effect of hypothermia and cardiac arrest on outcome of near-drowning accidents in children. J Pediatr 1990;117:179-83.

5. Gooden BA. Drowning and the diving reflex in man. Med J Aust 1972;2:583-7.

6. Bierens JJ, van der Velde EA. Submersion in the Netherlands: prognostic indicators and the results of resuscitation. Ann Emerg Med 1990;19(12):1390-5.

7. Ramey CA, Ramey DN, Hayward JS. Dive response of children in relation to cold-water near drowning. J Appl Physiol 1987;62(2):665-8.

8. Wong KC. Physiology and pharmacology of hypothermia. West J Med 1983;138(2):227-32.

9. Kamp AM and Sibert JR. Outcome in children who nearly drown—a British Isles study. Br Med J 1991;301:931-3.

10. Conn AW, Edmonds JF, Barker GA. Cerebral resuscitation in near drowning. Pediat Clin North Am 1979;26(3):691-701.

Documenting the Paper

CITING SOURCES IN THE TEXT

■ Acknowledge the source of all material that is not your own.

The text of a biological paper usually contains numerous *literature citations*, or references to the published studies of other authors. This is because scientists rarely work in a vacuum: hypotheses are developed, tested, and evaluated in the context of what other scientists have written and discovered. Most readers of biological literature are interested not only in the specific study being described, but also in the bigger picture — how particular findings contribute to our current understanding of an important scientific problem. Thus, careful *documentation*, or acknowledgment of the work of others, is essential to good scientific writing.

Biologists also need to provide literature citations because, like other writers, they have an ethical and legal obligation to give credit to others for material that is not their own. Such material includes not only direct quotations, but also findings or ideas that stem from the work of someone else.

■ Cite sources using a conventional format accepted by biologists.

Citation formats vary from one academic field to another. Unlike writers in the humanities, for example, scientists rarely use footnotes or endnotes to acknowledge sources. Instead, they insert citations directly in the text, either by giving the last name of the author(s) and publication date (*name-year system*), or by referring to each source by a number (*number system*). Complete bibliographic information is listed for each cited reference in the Literature Cited or References section of the paper. Each citation format has its advantages and disadvantages, and certain subdisciplines of biology (and each biological journal) tend to favor one method or the other.

A major advantage of the name-year system is that it is more informative. Many readers of biological literature are familiar with the major works in a particular field and will recognize well-known papers that are cited by the writer. In addition, the date of a particular study may be important to mention if its findings have been confirmed (or challenged) by later research.

The number system, however, conveys no information aside from the reference number. Readers who want to know who did the study, and when, must track down this reference in the Literature Cited section, unless the writer has supplied additional information in the text. Also, simple typographical errors — for example, typing the number 13 instead of 14 — can easily introduce inaccuracies. If a citation is added at the last minute, or if a citation is deleted, the writer must spend time renumbering all the references. However, some people find the number system less cumbersome and distracting, especially if the text contains a large number of citations. Therefore, this format is often used for review papers (see Chapter 5).

The following citation guidelines are based on recommendations by the Council of Biology Editors (recently renamed the Council of Science Editors, or CSE <http://www.cbe.org>) as given in the sixth edition of their handbook, *Scientific Style and Format: The CBE Manual for Authors, Editors, and Publishers* (1994). These guidelines are followed closely by some scientific journals, and with minor to major variations by others. Which citation method should you use? If you are preparing a paper for publication, you should follow the guidelines of the particular journal you hope will accept the paper. These are often titled "Instructions to Authors," and they usually appear somewhere in the journal itself. Follow the instructions carefully, down to the last detail. If you are writing a paper for a course, first ask your instructor which documentation method he or she prefers. If you are free to choose, then adopt one of the two CBE formats discussed below, or use the documentation style of one of your major references as a model. If you use one of your references as a model, study the way in which *its* documentation format has conformed to the requirements of the journal in which it appears. This is good practice for the time when you may be submitting your own work for publication. Whichever method you choose, be consistent and meticulous in adhering to the conventions. Such rules, even if they seem arbitrary, make the reporting of references an orderly activity, minimizing confusion for writers, readers, editors, and publishers.

Name-Year (Harvard) System

Cite each reference by giving the name(s) of the author(s) followed by the year in which the material was published.

WORK BY ONE AUTHOR

For each citation, use parentheses to enclose the name and the date. The name is followed by a space, but no comma:

> The most recent study of coral bleaching in this
> region (Scipione 1995) suggests that . . .

If the author's name appears as part of the sentence, put just the date in parentheses:

> In Chandler's (1993) study of marine planktonic
> ciliates . . .
>
> Black-horned locusts were first reported in
> Maryland by Rampolla (1980).

Because a name-year citation is considered part of the text, the period should always *follow* a citation that appears at the end of your sentence:

INCORRECT	The presence of hairs on a leaf can increase light reflectance from the leaf surface. (Cunningham 1966)
INCORRECT	. . . leaf surface (Cunningham 1966.)
CORRECT	. . . leaf surface (Cunningham 1966).

Variation: Some biological authors insert a comma between the name and the publication date: (Rampolla, 1980). This format (author's name plus comma and date) is also used by the American Psychological Association (APA), whose standard guidelines are followed by many writers in psychology, nursing, and other behavioral and social science fields.

WORK BY TWO AUTHORS

Put the senior author's name first. The senior author is the one whose name appears first after the title. It is generally assumed that he or she has had major responsibility for writing the paper.

> Other researchers (Silsby and Dunkle 1981) have
> suggested a different method of chemical analy-
> sis.
>
> Hagihara and Inoue (1993) found that . . .

Variation: Some journals use the ampersand (&) instead of "and":

> (Silsby & Dunkle 1981).

WORK BY THREE OR MORE AUTHORS

Cite the senior author's name followed by "and others," and then the date:

> White-lined bark beetles are attracted to the
> odor of rotting wood (Zorn and others 1992).
>
> Henry and others (2000) have suggested
> that . . .

Note the use of *have* (not *has*) in the example above. The subject (Henry and others) is plural and therefore must have a plural verb.

Variations: Instead of "and others," some authors use the Latin phrase "et alia" (abbreviated "et al."), either in plain text or italicized:

> Dovidio et al. (1995) found that . . .

Some journals specify that *all* authors be mentioned in the first citation but permit a shortened form of the same citation the next time the work is mentioned.

Two or More Works by the Same Author

Give the author's name followed by the dates of the works in chronological order. Use a comma followed by a space to separate the dates:

> The circulatory system of this species has been
> described in detail by Wylie (1978, 1980, 1983).
>
> Recent work on reaggregation of sponge cells
> (Zangrilli 1994, 1995) has shown that . . .

Two or More Works by the Same Author in the Same Year

Use letters (a, b, and so on) to differentiate between references. Retain this designation when listing them in the Literature Cited section.

> Rush (2000a, 2000b) postulated . . .

Variation: Some authors do not repeat the year:

> Rush (2000a, b) postulated . . .

Two or More Works by Different Authors

To cite more than one reference within the same parentheses, list them in chronological order, with the earliest first, and separate them using a semicolon followed by a space. To separate works by the *same* author, use a comma:

> Many different models have been proposed to ac-
> count for this phenomenon (Watson 1977, 1979;
> Kim 1988; Cox 1992).

If two or more sources are by different authors but published in the same year, put them in alphabetical order. In this example, Carter precedes Kim:

```
Many different models have been proposed to ac-
count for this phenomenon (Watson 1977, 1979;
Carter 1988; Kim 1988; Cox 1992).
```

AUTHORS WITH THE SAME LAST NAME

Occasionally two works published in the same year will have authors with the same last name. In this case, give the initials of each author in the citation:

```
Similar findings were reported for Colorado pop-
ulations of this species (Rider R 1990; Rider WA
1990).
```

WORK WITH AN ORGANIZATION AS AUTHOR

Give the name of the organization or group, followed by the year of publication.

```
Guidelines for evaluating dragonfly habitats as
potential conservation areas have recently been
published (International Committee for Preserva-
tion of Odonata 1995).
```

You may abbreviate organization names if they are long, or if you need to cite them frequently:

```
Guidelines for evaluating dragonfly habitats as
potential conservation areas have recently been
published (ICPO 1995).
```

WORK WITH NO IDENTIFIABLE AUTHOR

If a source has no identified author, you may refer to the author as "Anonymous":

```
This species has been reported in both Madison
and Oneida Counties of New York (Anonymous 1986).
```

Variation: Some authors cite a source with no known author by using the title instead of the word "Anonymous":

```
Similar employment trends were noted for the
following decade (Report of the Task Force on
Opportunities for Women in Pediatrics 1983).
```

In this case, the first important word in the title is used to alphabetize the source in the Literature Cited section.

ARTICLE OR CHAPTER IN AN EDITED VOLUME

To cite a reference in an edited collection of works by different authors, you need not give the editor's name. Just give the name of the author(s) of the particular article or chapter, along with the year the edited collection was published. For example, suppose that one of your sources is "Sexual Selection and Mate Choice," by Timothy R. Halliday, which appears on pages 180–213 in the book *Behavioural Ecology: An Evolutionary Approach*, edited by J. R. Krebs and N. B. Davies, and published in 1978. Your in-text citation would be (Halliday 1978).

A SPECIFIC PART OF A WORK

If you need to refer to a particular table, figure, or other portion of a source, include this information within the parentheses, as shown:

```
Simmons and Grimaldi (1955, Table 4) found
that . . .

Similar data were obtained by Linsley (1973,
Figure 2).
```

Note the use of a comma, followed by a space, after the date.

A DIRECT QUOTATION

Some authors give the page number in parentheses, separated from the date by a comma. Others supply just the author and date:

```
Rith-Najarian (1987, p 3) has written, ". . .

According to Rith-Najarian (1987), ". . .
```

UNPUBLISHED MATERIAL

Occasionally you may need to cite a personal conversation with someone, an informal written communication, or the unpublished data of a colleague. E-mail correspondence — for example, with colleagues or with a known expert in a particular field — also falls into this category. Professional ethics dictate that you obtain the person's permission to include such material in your paper. For the citation, use the last name and initials of the person who supplied the information, followed by a word or two of explanation, either written out in full or abbreviated. Do not list these unpublished materials in the Literature Cited section because they are inaccessible to your readers. CBE format requires that you note that such citations are not listed with your other references.

```
D. Craine (personal communication; unreferenced)
has suggested that . . .

This species has also been found along the
shores of lakes and ponds (L Arnold, 2000, un-
published data; unreferenced).
```

If you refer to your *own* unpublished data from a *different* study, give your name followed by "unpublished" or "unpub."

Professional biologists are often familiar with their colleagues' recent manuscripts and may refer to them in their own writing. When an article or book has been accepted for publication but has not yet been printed, it is regarded as "in press" or "forthcoming." (Recent CBE guidelines prefer the term "forthcoming," because it can be used not only for printed materials but also for electronic sources.) When a source is in press or forthcoming, indicate this in the citation:

```
Similar data were collected by Gelder (in
press).

Wilson and Joseph (forthcoming) found
that . . .
```

If you are certain about the future publication date for a particular source, you may give this date in the citation and then note in the Literature Cited section that the source is forthcoming.

Variation: Some journals do not permit citations of unpublished observations, personal communications, or manuscripts in preparation.

WORKS YOU HAVE NOT CONSULTED DIRECTLY

Avoid referring to sources you have not read yourself; the reader assumes that you have firsthand knowledge of all the works you have cited. Occasionally, however, you may need to cite an important source that is not accessible to you. If so, specify where you acquired your secondhand knowledge of this source:

```
Arcari (1954, cited in Castrodale 1985) reported
that . . .
```

List *both* sources in your Literature Cited section.

COURSE HANDOUTS OR LABORATORY MANUALS

For academic assignments, you may wish to cite an unpublished handout or lab manual that is specific to a particular course. One way to do this, if you are using the name-year citation system, is to give your instructor's name (as author) along with the year:

```
Sampling methods followed the procedure outlined
by Brown (1999).
```

Then list the lab manual alphabetically under the author's last name in your Literature Cited section, followed by the year, title, your school's location and name, and the number of pages. Alternatively, if you are using the number system for documentation, you can refer to the lab manual by number, as you would with any other source, and again give full information about it in your Literature Cited section (see pp. 117–25).

If there is no clear author of the lab manual (as in the case of a frequently offered course by a changing team of instructors), you could give the manual title followed by the appropriate year:

```
Community similarity was determined using a mod-
ification of Morisita's Index (Ecology 204 Labo-
ratory Manual 1995).
```

List the manual in your Literature Cited section alphabetically, using the first word of the title in place of author, and adding the location and name of your institution, as well as the number of pages.

Finally, your instructor may have his or her own other preferences about citing such sources.

Number System

In this system, each source is given a number, which appears as a superscript at the appropriate place in the text:

```
Sap pressures in Douglas fir[1] become more nega-
tive as relative humidity rises.
```

If you mention a particular source more than once, use the same number originally assigned to it. CBE format requires that sources be numbered in the order in which you *cite* them in your paper. They are then listed in *numerical order* in the Literature Cited section. (This method is called the "citation-sequence system" in the CBE manual.) If you cite two or more sources at the same point in the text, use commas (with no intervening spaces) to separate the numbers. If three or more of your reference numbers are consecutive, use an en dash (or a hyphen) to indicate this:

```
Plantago major is common in heavily trampled
areas, such as the edges of roads and
sidewalks.[2,3]

Cytological studies of development from dormant
gemmules[4,6-8] have shown that . . .
```

As in the name-year system, if you have not consulted a particular source directly, you need to indicate that you are relying on secondhand information:

```
The model of water exchange proposed by Pinchin
and Muller[11] (cited in 12) . . .
```

If the author or publication date of a particular study is important to your discussion, add this information to the sentence:

```
Fuller[14], studying three species of caddisflies
in Montana, was the first to observe that . . .
```

```
This species was not listed in early floras of
New York; however, in 1985 it was reported in a
botanical survey of Chenango County.⁹
```

Variations: Some authors place citation numbers in parentheses and on the same line as the text. The numbers are separated by commas, with a space after each comma:

```
Previous studies (3, 4) indicated that . . .
```

Some journals using the number system specify that sources be numbered in *alphabetical order* and listed alphabetically in the Literature Cited section.

■ Put citations where they make the most sense.

Whichever documentation system you use, put each citation close to the information you wish to acknowledge. Do not automatically put cited material at the end of every sentence. The following statement, for example, is ambiguous:

```
Pollination of Linaria vulgaris has been studied
in both the field and the laboratory (Arnold
1962; Howard 1979).
```

Did Arnold do his studies in the field and Howard in the laboratory? Or Howard in the field and Arnold in the laboratory? Or both authors in both settings? Moving the first citation clarifies the situation:

```
Pollination of Linaria vulgaris has been studied
in both the field (Arnold 1962) and the labora-
tory (Howard 1979).
```

If you actually meant to say that both Arnold and Howard did both types of studies, then you are better off rewording the sentence:

```
Arnold (1962) and Howard (1979) have studied the
pollination of Linaria vulgaris in both the
field and the laboratory.
```

■ Do not cite sources for information regarded as common knowledge in a particular field.

Although it is not strictly wrong to do this, it is unnecessary. If certain material is well known and fundamental to a particular field, you need not cite sources. For example, you need not cite your textbook or other references to say that living organisms are composed of cells or that meiosis in higher animals gives rise to haploid gametes. Such material is general knowledge that is familiar to the audience of any biological paper. Similarly, any subdiscipline of biology has information that is regarded as elementary and basic by anyone working in that field.

How do you decide what is common knowledge and what is not? This ability comes with experience as you grow more familiar with a subject and the literature on it. For academic assignments, the background knowledge you share with your classmates will be your best guide. If in doubt, ask someone more experienced than yourself, or cite the source anyway.

■ Use citations carefully.

Citations allow you to acknowledge the work of others; they also *inform* the reader. Do not pack your text with citations simply to demonstrate that you've done your homework and are intimately familiar with the literature. Consider the following example:

```
Many studies have been made of the factors
influencing variable mate-guarding in the
dragonfly Plathemis lydia (Brubaker 1979; Darby
1980; Kraly 1981; Jamieson, Pegg, and Joy 1983;
Angell 1984; Napolin 1985; Hayes 1988; Gittens
1991).
```

In this passage, vague reference to a large number of works does little to enlighten the reader. Decide which of your many sources are *most* important and why. Refer to them in a meaningful way so that they promote a focused discussion:

```
Many studies have been made of the factors in-
fluencing variable mate-guarding in the dragon-
fly Plathemis lydia. Brubaker (1979) found
that . . . , whereas Darby (1980) showed
that . . . Experimental manipulations of male
density at breeding sites by Angell (1984) and
Hayes (1988) suggested that . . .
```

Watch out for unintentional sexism when you refer to authors. Do not automatically assume that every biologist is male:

```
Virkler (1995) investigated bacterial resistance
to the antibiotic chloramphenicol. She suggested
that . . .
```

■ Avoid unnecessary repetition of citations when the context is clear.

When using the name-year system, you need not repeat the author's name if it already appears earlier in the sentence and you are not citing additional authors:

```
Huerta (1960) conducted the first laboratory
studies of this phenomenon, and reported in
later works (1970, 1977) that . . .
```

When using either the name-year or the number system, you need not repeat the same citation in every sentence when you are discussing the work at some length. Again, let the need for clarity be your guide. In the following example, note that the writer has cited the source at the start of her paragraph and again, near the end, to remind the reader that she is still referring to the same study:

```
Nakhimovsky and Ochs (1994) studied interspe-
cific interactions between two species of libel-
lulid dragonflies (Libellula pulchella and
Plathemis lydia) at a small pond in upstate New
York. They found much overlap between the two
species in the use of shoreline areas for pair
formation, territoriality, mating, and oviposi-
tion. Interspecific interactions between males
seeking perch sites were relatively infrequent;
however, male P. lydia often disrupted oviposi-
tion attempts by L. pulchella females. Nakhi-
movsky and Ochs (1994) suggest several promising
lines of research on sex recognition in these
two species, which are characterized by strik-
ingly similar females.
```

PREPARING THE LITERATURE CITED SECTION

■ Understand the difference between a Literature Cited section and a bibliography.

A bibliography contains all the sources mentioned in the text, along with additional references on the topic. The Literature Cited section contains *only* the sources that have been *cited* (referred to) in the text. Even if you have acquired useful background knowledge by reading five articles and three books, do not list any of them in the Literature Cited section unless you have specifically mentioned them in your text. Bibliographies are not generally part of scientific papers.

The Literature Cited section is also commonly titled References, or sometimes References Cited.

■ Report sources completely and accurately.

After typing up the main body of the paper, you may be tempted to race through the listing of references, assuming that no one looks at this part anyway. Do not underestimate the importance of a meticulously prepared Literature Cited section. Readers do look at this section. In their minds, the amount of attention you have given it reflects the care given to the rest of the paper. If your sources are reported sloppily, people may doubt your authority, integrity, and thoroughness as a researcher.

The Literature Cited section also serves as an important source of references for readers who want further information on the topic. You owe these people accuracy and completeness. Even some published papers contain mistakes or are missing information in the list of references. Few things are more frustrating for a researcher than to be told the wrong page numbers or journal volume for an article he or she needs to track down.

Long *before* you sit down to prepare the Literature Cited section, become familiar with the kinds of bibliographic details you will need for each type of source (paper, book, article in an edited collection of articles, Web document, and so on). As you read and take notes, record all this information in a master list or on file cards. (See pp. 16–19 for more on note-taking techniques.) Doing so will save you time and avoid chaos later.

■ Use a conventional format for listing your sources.

As with in-text citations, biological journals vary with respect to the format used for the Literature Cited section. In fact, you won't have to plunge too deeply into the primary literature to discover a bewildering array of variations from journal to journal in this respect. For example, consider the following entries taken from the Literature Cited section of papers appearing in 1995 in four different journals:

LEE, T. D., and F. A. BAZZAZ. 1982. Regulation of fruit and seed production in an annual legume, *Cassia fasciculata*. Ecology 63: 1363–1373. (*American Journal of Botany*)

1. Sen, R. & Baltimore, D. *Cell* **47**, 921–928 (1986). (*Nature*)

[1] Auerbach, S., Zhou, F.C., Jacobs, B.L., and Azmitia, E., Serotonin turnover in raphe neurons transplanted into rat hip-

pocampus, *Neurosci. Lett.*, 61 (1985) 147–152. (*Developmental Brain Research*)

Low, M. G., and Saltiel, A. R. (1988). Structural and functional roles of glycosylphosphatidylinositol protein anchors. Science *239,* 268–275. (*Cell*)

Prospective authors prepare the Literature Cited section by closely following the guidelines prescribed by the journal for which they are writing. For academic assignments, check if your instructor has certain format preferences; otherwise, use one of the standardized CBE formats below or follow the system used by a recent paper on your topic.

The following examples are based on recent CBE guidelines for listing references using either the name-year citation system or the number system. These examples should help you with most of the references you're likely to encounter as you begin biological research. See the CBE manual or other resources for information on citing specialized technical reports, government documents, audiovisual publications, newspaper articles, and other sources not covered here.

JOURNAL ARTICLE WITH SINGLE AUTHOR

NAME-YEAR SYSTEM	Sachar DB. 1994. Budesonide for inflammatory bowel disease: is it a magic bullet? N Engl J Med 331:873-4.
NUMBER SYSTEM	36. Sachar DB. Budesonide for inflammatory bowel disease: is it a magic bullet? N Engl J Med 1994;331:873-4.

These examples show how to list the same paper in the Literature Cited section using each of the two citation methods. You will notice many similarities in format. The last name of the author is followed by his or her *initials* only, and these are not separated by periods or spaces. Only the first word of the paper title is capitalized. The journal title (here, *The New England Journal of Medicine*) is capitalized and abbreviated, but it is not underlined or italicized. To find the conventional abbreviation for a journal, look at a copy of the journal itself, or consult your reference librarian for help. If the title is a single word, or if it has no conventional abbreviation, write it out in full. Also note that some journals do not allow journal abbreviations of any kind.

In the name-year system, the year of publication follows the author's name. After the year comes a period followed by the title of the paper, the journal title, the volume of the journal, a colon, and the pages on which the paper appears.

The number preceding the author's last name (36) is the citation number used by the writer to refer to this source in the text. In the number system, the year of publication follows the journal title. After the year comes a semicolon, followed by the volume of the journal, a colon, and then the pages on which the paper appears. Note the absence of spaces between these components of the reference listing. You will notice these two differences (in the placement of the date and the presence or absence of the semicolon) in the following examples of other types of reference listings.

For journals that number each issue separately (not continuously through a single volume), it is customary to give the *issue number* after the volume number, in parentheses. This practice reduces confusion and helps the reader find the appropriate issue more efficiently.

NAME-YEAR   . . . J Fam Pract 29(6):665-70.
SYSTEM

NUMBER   . . . J Fam Pract 1989;29(6):665-70.
SYSTEM

JOURNAL ARTICLE WITH TWO OR MORE AUTHORS

The first, or senior, author is listed first, followed by the coauthor(s) in the order in which they appear on the title page:

NAME-YEAR   Binder V, Hendriksen C, Kreiner S.
SYSTEM   1985. Prognosis in Crohn's disease—
             based on results from a regional pa-
             tient group from the country of Copen-
             hagen. Gut 26:146-50.

NUMBER   18. Kozlowski RD, Steinbrunner JV,
SYSTEM   MacKenzie AH, Clough JD, Wilke WS,
             Segal AM. Outcome of first-trimester
             exposure to low-dose methotrexate in
             eight patients with rheumatic disease.
             Am J Med 1990;88:589-92.

In the CBE format, the names of *all* authors of a paper (up to a maximum of 10) are given in the Literature Cited section. If there are more than 10 authors, list the first 10, followed by "and others."

Note that the single-word journal title in the name-year system example is not abbreviated. Note also that proper nouns (Crohn's, Copenhagen) are capitalized, along with the first word of the paper title.

BOOK

The rules for authors and titles of books are similar to those for journal articles. Include the edition if there is more than one edition of a book. Add the designation "editor(s)" if applicable. The place of publication precedes the publisher, separated by a colon. The total number of pages follows at the end.

NAME-YEAR SYSTEM	Danforth DN, editor. 1982. Obstetrics and gynecology. 4th ed. Philadelphia: Harper and Row. 1316 p.
NUMBER SYSTEM	6. Sokal RR, Rohlf FJ. Biometry: the principles and practice of statistics in biological research. 2nd ed. San Francisco: WH Freeman; 1981. 859 p.

ARTICLE OR CHAPTER IN AN EDITED VOLUME

NAME-YEAR SYSTEM	Petter JJ. 1965. The lemurs of Madagascar. In: DeVore I, editor. Primate behavior: field studies of monkeys and apes. New York: Holt, Rinehart and Winston. p 292-319.
NUMBER SYSTEM	9. Halliday TR. Sexual selection and mate choice. In: Krebs JR, Davies NB, editors. Behavioural ecology: an evolutionary approach. Oxford: Blackwell Sci Pub; 1978. p 180-213.

In the examples above, notice that the reference listing begins with the name of the author of the particular article or chapter, not with the editor(s) of the whole collection. Be sure to include the page numbers (p 292–319) of the section you have cited. Also, note the British spelling of the word "behavioural" (as opposed to the American "behavioral") in the second reference. Always retain the original spelling used by any source you use.

WORK WITH ANONYMOUS AUTHOR OR ORGANIZATION AS AUTHOR

NAME-YEAR SYSTEM	[ICPO] International Committee for Preservation of Odonata. 1995. Guidelines for evaluating dragonfly habitats as conservation areas. Boston: Entomological Pub Co; 12 p.
NUMBER SYSTEM	24. [Anonymous]. Growth records for organ pipe cactus in the Ajo area. Desert Life 1971;6:2-3.

In the name-year system example, note that the abbreviation used for the in-text citation precedes the full name of the organization and is placed in brackets.

FORTHCOMING WORK

The format for articles or books that are soon to be published is similar to that for published works. Include the expected date of publication, if known.

NAME-YEAR SYSTEM	China-lai V. 2002. Mosses of Long Island. Shoreham (NY): Shoreham Univ. Pr. Forthcoming.
NUMBER SYSTEM	5. Johnson MR. Dietary studies of female office workers. J Nutr Ed. Forthcoming.

THESIS OR DISSERTATION

NAME-YEAR SYSTEM	Grapard U. 1982. The floral biology of *Linaria vulgaris* (Scrophulariaceae) [dissertation]. Westbury (NY): Phipps College. 101 p.
NUMBER SYSTEM	3. Arnold KM. Feeding preferences in the spiny lizard (*Sceloporus jarrovi*) [MSc thesis]. Tucson (AZ): Southwestern University; 1979. 77 p.

If you have additional information about the availability of a dissertation (for example, through University Microfilms in Ann Arbor, Michigan), this information can be added after the number of pages.

CONFERENCE PAPERS

In the name-year system, give the name of the author, year of publication, paper title, name of the conference, dates and place of the conference, place of publication, publisher of the proceedings, and the page numbers for the paper:

Rith J. 1988. Plant succession on abandoned railways in rural New York State. In: King, T. editor. Northeastern Ecology Society VI Symposium; 1988 May 2-5; Syracuse, NY. Boston: Northeast Pr. p 34-41.

The CBE format for the same reference using the number system is similar, except that the publication date is given after the name of the publisher:

16. Rith, J. Plant succession . . . Boston: Northeast Pr; 1988. p 34-41.

ELECTRONIC SOURCES

To cite a paper in an electronic journal (that is, a journal with no print version), CBE recommends the following formula for the name-year system: Author(s)' last name(s) followed by first initial(s), publication date, paper title, journal title (conventional abbreviation), volume and issue numbers, page numbers of paper, availability information, and date you accessed the paper. After the journal title, you also need to indicate in brackets that the journal is an "online serial."

The format for the number system is similar, except that the publication date is placed after the name of the journal.

NAME-YEAR SYSTEM	Kingsolver JC, Srygley RB. 2000. Experimental analyses of body size, flight and survival in pierid butterflies. Evol. Ecol. Res.[serial online];2:593-612. Available from: Colgate University online catalog. Accessed 2000 Oct 3.
NUMBER SYSTEM	11. Kingsolver JC, Srygley RB. Experimental analyses of body size, flight and survival in pierid butterflies. Evol. Ecol. Res.[serial online] 2000;2:593-612. Available from: Colgate University online catalog. Accessed 2000 Oct 3.

Note that papers in some electronic journals may be identified not by volume, issue, and page numbers but instead by the day, month, and year published and a document number. For such papers, give this information instead.

To cite a personal or professional Web site, list the name of the author (or organization) along with the publication date (or date last revised), title of Web site, URL in angle brackets, and the date you accessed the site:

[CDC] Centers for Disease Control and Prevention. 2000 Oct. 2. CDC home page. <http://www.cdc.gov.>. Accessed 2000 Oct 3.

The date of access is important to include because Web sites and other resources on the Internet tend not to be permanent documents: their addresses and even their content are subject to sudden change. A Web site that is here today may be gone tomorrow. Consequently, it is important to report when this information was available to you. Also, be sure to record the Web address carefully: the URL must be reproduced exactly as it appears in order for another person to locate this site efficiently. For more suggestions about how to cite electronic sources, see *Online! A Reference Guide to Using Internet Sources,* by Andrew Harnack and Eugene Kleppinger (Bedford/St. Martin's, Boston, 2000). Your instructor may also have his or her own preferences.

Finally, remember to use discretion in including electronic sources as important references in your paper (see Chapter 1). Internet resources vary greatly in quality and authority, as opposed to the more established, peer-reviewed primary literature of scholarly journals, much of which may not be available to you electronically. Although the Internet may be highly useful during the early stages of your research — while you're still brainstorming for ideas, gathering background (secondary) information, and tracking down prominent researchers — Web sites and the like should not represent the majority of your references. Check with your instructor, who may prefer that you list only certain kinds of Internet sources — for example, important government databases or primary articles in electronic journals — in your Literature Cited section.

■ Arrange references either alphabetically or numerically.

If you have used the *name-year system,* remember that your references should be listed *alphabetically* according to the senior author's last name. (See the sample research paper in Chapter 4 for an example of this method.) If a particular author has published more than one work, list them in chronological order beginning with the earliest. If you have used letters to distinguish between papers by the same author(s) published in the same year (for instance, Swarthout 1947a, Swarthout 1947b), then be sure to retain these designations, listing the 1947a paper before the 1947b one. Works by two or more different authors with the same last name are arranged alphabetically according to the authors' first initials. For example, an article by E. J. Arnold would come before one by L. Arnold.

If you have used the *number system* (or *citation-sequence system* as in the CBE style manual), remember that you have two options here: either (a) number and list the references in the order in which you *cite* them in the paper, or (b) list the references *alphabetically*, then number them, and use these numbers as the citation numbers in the text. Current CBE guidelines use the first method, but your instructor (or the journal you are writing for) may prefer the second one.

Journals also vary with respect to the use of indentation in the Literature Cited section. For academic assignments, it is usually acceptable to begin the first line of each entry at the left margin, and then indent five spaces for successive lines (see sample paper, Chapter 4).

■ Proofread your Literature Cited section for accuracy, completeness, and consistency.

Whichever documentation format you use, every entry must conform as closely as possible to the guidelines. Check that every source you cite is actually *listed* in the Literature Cited section. Check that no source appears in the Literature Cited section that is not *cited* in your text. You must also focus on minutiae, scrutinizing with meticulous care every punctuation mark, every space, every number, every capital letter. The task may seem tedious indeed. Just remember that all this time and effort is indispensable to the production of a well-received paper.

CHAPTER 7

Drafting and Revising

Expect to write several drafts of your paper before you are satisfied with the final product. Good writing is generally the product of careful *rewriting,* or *revising,* in which you evaluate your early attempts at organizing and expressing your ideas. In the process you end up scrutinizing the ideas themselves, as well as your own mastery of the subject. You may find that until you can express a concept clearly enough that others can understand it, you have not fully understood it yourself. Working through successive drafts also teases out *new* ideas, perhaps connections between disparate sets of data or new insights about the significance of your research. As you grope for more appropriate wording and a clearer structure, you give form and substance to thoughts still lurking in your subconscious. Gradually and sometimes painfully, you discover what it is you really want to say.

THE FIRST DRAFT

Many people assign too much importance to a first draft. Think of it as just a rough version of the paper, an exercise in organizing your thoughts and getting down the main ideas. Focus on content, not on prose style or mechanics. Because much of what you write now may not end up in the final version, it is unproductive to tinker with commas, spelling, or sentence construction. First get down the basic information; in later drafts you can look more closely at your writing style and such matters as grammar and punctuation.

How do you shift from doing research or taking notes to composing the first draft of the paper? Experienced writers use a variety of methods. The following ones may help you get started.

■ Devise a working title.

The title should reflect your most important findings and your purpose in writing the paper. What is the main point you wish to make? What new information or perspective does your study contribute? Deciding on a preliminary title early in the writing process helps you focus your thoughts and start drafting the paper. If necessary, you can revise the title later as your ideas take shape.

■ Make an outline or a rough plan.

All writers need some sense of how the various points they want to make are logically connected to one another. However, people vary tremendously with respect to how organized their thoughts must be before they start to write. Some writers find that a detailed, point-by-point outline is absolutely essential before they start the first draft. Others prefer to work out a more general framework that leaves them considerable flexibility. Some writers find any kind of outline confining. Instead, they plunge directly from their notes into the first draft, doing most of the organizing in their heads as they go along.

Systemic planning may be more appropriate for some sections of the paper than for others. For example, the Materials and Methods section of a research paper is fairly straightforward and contains many details that must be organized in a logical manner. Once you plot it out carefully, it may practically write itself. The same may be true of the body of many review papers. By contrast, the complexities of a Discussion section may be difficult to pin down and categorize. In fact, you may not be sure of what you want to say until you start putting words to paper and have worked through one or two drafts.

It pays to experiment with different forms of planning. If you don't like outlines, try jotting down a rough list of the major topics or issues you need to discuss. You may find it easier to visualize the relationships between ideas if you make a "flow chart" with lines, arrows, or brackets linking groups of words or phrases. If you do rely on detailed outlines, remember that they should not be inflexible schedules. They can be stretched or restructured to conform with your writing as it develops. An outline can be a valuable tool to get you started, but it should never be an end in itself. It should never keep you very long from that most important act of plunging into the paper.

■ Start the easiest writing first.

Don't feel you have to begin at the beginning. If you are writing a research paper, for example, it may not be productive to start with the Abstract, or even with the Introduction. Begin with whatever part seems easiest and most straightforward. For many people this is the Materials and

Methods section, which often can be drafted before the research is completed. The Results section also may be relatively easy to put together. Generally, the Abstract is best written last, after all other major sections of the paper have been completed. The point is that once you start drafting *any* section of the paper, you have broken the barrier between you and the material, and the writing will gain momentum.

■ Talk to others.

Brainstorming with other people about the project may help get you started by giving you an audience on which to test out your ideas. Professional scientists rely heavily on such feedback from their colleagues, both in the more formal settings provided by conferences and on a day-to-day basis during lunch or coffee breaks. Electronic mail and Internet discussion groups also provide a forum for informal conversation. Intellectual exchange with others helps scientists keep their work on track and instills it with fresh insights. If you read the Acknowledgments portion of several scientific papers, you will see how important such discussions are to research and writing processes.

PRACTICAL SUGGESTIONS FOR REVISING

Revising will be most efficient if you break the task down into several successive steps. First, consider the paper as a whole, checking the overall content, organization, coherence, and consistency of your argument. Try to resist the impulse to make small-scale changes — for example, changes in punctuation — until you are satisfied with the main substance and structure of your paper. This is particularly important when you are composing on the computer. Because word processing makes it so simple to alter the surface features of your text, you may be tempted to make even your first draft "perfect," with every comma in place and every typographical error corrected. However, doing so will distract you from the more important job of articulating and organizing your main ideas.

Once you are satisfied with the paper's content and general organization, work on polishing your prose style, examining individual paragraphs and sentences closely for clarity, accuracy, and conciseness. Look for wordiness, awkward sentence constructions, improper word usage, faulty subject-verb agreement, excessive use of the passive voice, and other problems that may obscure your meaning and impede the flow of your writing. Finally, proofread the text for punctuation, spelling, and typographical errors, and inspect the final form of the manuscript.

In practice, of course, making revisions is not quite so orderly and mechanical a process. Also, "writing" and "revising" a paper are not totally separate activities: writers typically make changes as they are composing, not just after they have written a draft all the way through. Nevertheless, your writing may be more productive if you set some priorities for yourself, focusing on big changes first, small changes later. Becoming more aware of how you write will help you develop the drafting and revising strategies that work best for you.

Some writers work best if they do their early composing in longhand and save the later stages of revising and proofreading for the word processor. Others need the familiar hum of the computer right from the start and prefer to see their writing appear neatly on the screen in front of them. If you do write out early drafts in longhand, the "cut-and-paste" method saves time and effort when you need to move chunks of text from one place to another. Use scissors and tape to patch the new version together, rather than write everything over again. Of course, word processing allows writers to move around words, sentences, and sections of text and to make sweeping changes in the general structure of the manuscript.

Do not discard your early drafts; keep everything at least until you have submitted the final version of the paper. Sometimes parts of a first or second draft end up sounding better than the corresponding sections of later drafts. Put both the page number and the number of the draft on each piece of writing in case you mix up parts of different drafts.

Do not confine all your revising to the computer screen. The computer shows you only a small part of the manuscript at one time, but some problems may be more noticeable when you see several pages of the paper spread out in front of you. For this reason, it makes sense to print your work frequently and make additional changes by hand directly on the manuscript. Keep all the printouts even after you have incorporated the changes in your file. You may need to refer to an early draft before you are finished with the paper.

During the later stages of editing and proofreading, remember to use word processing features that will save you time. For example, "search and replace" features of word processing programs are invaluable for locating specific words or phrases in the text and allowing you to correct misspellings or other errors. Suppose you realize that throughout the paper you have typed the genus *Gomphus* incorrectly as *Gomphis*. On command the computer can locate every place where *Gomphis* appears in the text and can automatically make the necessary correction.

Spelling checkers will locate misspelled words or typos such as *reserch* instead of *research*, or *studyy* instead of *study*. However, they cannot detect typographical errors that are legitimate words, such as *samples* instead of *sample*. They also can't distinguish between the various spellings of words such as *two, to,* and *too;* or *principle* and *principal*. Finally, although spelling checkers have sizeable dictionaries, they may not include many of the words you are using, such as specialized terms and proper nouns. (However, many spelling checkers allow you to enter such words into their dictionaries.) Therefore, do not rely on spelling checkers to catch all errors, and be sure to carefully proofread your final draft. (For more on proofreading, see p. 160.)

Allow plenty of time for revising and proofreading your paper. Set each draft aside for several hours or several days; when you return to your writing you'll view it more objectively and work more productively. Try reading a draft out loud; you will be amazed at how useful this technique is for spotting awkwardly worded or ungrammatical constructions. Ask one or two of your friends to criticize the manuscript. Pick people who will give honest and thoughtful opinions and will not just say what they think you want to hear. Professional biologists rely heavily on criticism from their colleagues as they prepare manuscripts. Once papers are submitted to journals they are subjected to further scrutiny by editors and outside reviewers. This review process, as formidable as it may sound, gives authors invaluable feedback about their work and helps to ensure high standards for research and writing.

Finally, but most importantly, make sure you have two copies of your file, either one on the hard disk and a backup on a diskette, or one copy on each of two different diskettes. (Store the diskettes in different places when you are not working on the computer.) There is always the possibility that you may lose or damage a diskette, or that you may accidentally erase your file, thus losing hours or days of hard work. Remember to update the backup file each time you make revisions.

CHECKING CONTENT AND STRUCTURE

■ Improve logic, continuity, and balance.

In a research paper you need to keep track of your argument. What major question are you addressing? What hypotheses have you attempted to test? Do the data you present in your Results directly relate to these hypotheses and to the conclusions put forth in the Discussion? A good scientific paper is somewhat like a mystery story with the "solution" well supported by a carefully crafted body of evidence in the Results section. However, the scientific writer has also gradually prepared the reader for the major conclusions, which should come as no sudden surprise.

Although a review paper has a less formal structure, it also needs to be built around a logical train of thought. Readers need to understand your rationale for choosing the topic. They need a sense of movement as you develop your main points, and they need to see the paper end on a satisfying note through summarizing statements and conclusions.

Once you have completed a rough draft, try to visualize the paper as a unit. Is its structure clear? Do you lead readers along your line of reasoning point by point, paragraph by paragraph, making clear transitions from one topic to the next? Or do they have to grope their way along, guessing the connections? Are any topics underdeveloped or overdeveloped, throwing the paper out of balance? Do you maintain a consistent style and approach? Compare your Introduction with the rest of the paper. Have you delivered to the reader what you promised at the beginning? Are your initial statements compatible with your concluding ones?

Sometimes large-scale revisions can be accomplished relatively effortlessly by moving sentences or even whole paragraphs from one place to another. Usually, however, you need to do substantial rewriting — paring down overwritten passages, modifying highly speculative comments, developing poorly expressed ideas, adding new clarifying material, and revising the beginnings and ends of paragraphs so that you can fit them into new contexts. Such revisions can be frustrating, even painful. However, they lead to a new product: a clearer, more organized version of the paper.

■ Omit unnecessary material.

Even when you feel you have "nothing to say," you may end up with more material than you anticipated, perhaps more than you really need. As you check the manuscript for logic and balance, be alert for places where you strayed off the topic. The Materials and Methods section may include procedures that do not pertain to any of the data discussed. The Results can easily become a grab bag of miscellaneous data, only some of which are important to the main story. The Discussion may be weighed down by predictions impossible to test in the near future or by rambling comments of only peripheral importance. Check that *every* point is relevant to your main objectives. Be ruthless. Cut any extraneous material — text, table, or figure — no matter how hard you worked on it.

■ Check for completeness and consistency.

Writing a research or a review paper requires you to keep track of many details. It is surprisingly easy to create inconsistencies or to leave out pieces of needed information. Here is a checklist of questions to ask yourself:

- Is the title specific and informative?
- Does the Abstract and/or Table of Contents include all relevant parts of the paper?
- Does the Introduction state your main objectives, hypothesis, or thesis?
- Have you included enough information in the Materials and Methods section to enable someone else to repeat your study?

- Have you explained in Materials and Methods the procedures for collecting all the data presented in the Results?
- Are figures and tables numbered consecutively in separate series?
- Is every figure and table cited correctly in the text?
- Do the data in each figure or table agree with your in-text discussion?
- Do any figures or tables present conflicting data?
- Are data in related figures or tables shown in a consistent manner?
- Is each table and figure understandable apart from the text?
- Are any important results missing?
- Have you used enough headings and subheadings to guide the reader?
- Does the Discussion address the major implications of your findings?
- Have you considered problems, inconsistent results, and counter-evidence?
- Have you cited all necessary sources?
- Are all sources cited in the text listed in the Literature Cited section?
- Does the Literature Cited section include any sources not cited in the text?

IMPROVING PARAGRAPHS

■ Present coherent units of thought.

Paragraphs are not just chunks of text; at their best, they are logically constructed passages organized around a central idea often expressed in a *topic sentence*. A writer constructs, orders, and connects paragraphs as a means of guiding the reader from one topic to the next, along a logical train of thought.

Topic sentences often occur at the beginning of a paragraph, followed by material that develops, illustrates, or supports the main point:

> *The teeth of carnivorous and herbivorous vertebrates are specialized for different ways of life.* Those of carnivores are adapted for capturing and subduing prey and for feeding largely on meat. Dogs and cats, for example, have long, sharp canines used for piercing, and molars and premolars equipped for cutting and tearing. By contrast, herbivores such as cows and horses have teeth specialized for feeding on tough plant material and breaking down the indigestible cellulose in plant cell walls. Their molars and premolars have large, ridged surfaces useful for chewing, gnawing, and grinding.

Do not distract the reader by cluttering up paragraphs with *irrelevant* information. For example, in the passage below, the second sentence does not relate to the paragraph's main point as stated in the topic sentence preceding it. We can strengthen the paragraph by deleting the second sentence altogether.

> Tanner (1981) sheds new light on the processes that may have been critical in the evolution of early hominids from chimpanzee-like ancestors. *Tanner also includes much interesting information about chimpanzee social life.* She suggests that a critical innovation might have been the extensive use of tools (initially, organic materials such as bones or sticks) by females for gathering. Such tools, she argues, would have been employed long before weapons were being manufactured by men for hunting large game.

How long should a paragraph be? There is no set rule; paragraph length depends on the writer's topic, coverage, format, and purpose. As a working rule, aim for four to six sentences, then shorten or lengthen the passage as needed. Avoid one-sentence paragraphs. Used skillfully, for example in informal essays or in fiction, they give variety and emphasis. However, they may express a poorly developed idea or one that really belongs in a neighboring paragraph. Conversely, very long paragraphs can become unwieldy and confusing. Usually such passages contain more than one idea and can be divided.

Keep in mind that it is not enough to group related sentences together into a paragraph. You must also *demonstrate* these relationships; otherwise, the reader cannot follow your line of thought. Look at the following passage from a lab report:

> Plants provide a constant supply of oxygen to our atmosphere. Both plants and animals depend on oxygen in the utilization of their food. During the process of photosynthesis, plants consume carbon dioxide and release oxygen.

The reader has to struggle to find the point of this paragraph. One problem is that the sentences seem to be thrown together haphazardly. Another is that the writer provides no conceptual links from one sentence to the next. Here is a revised version of the paragraph:

> During the process of photosynthesis, plants consume carbon dioxide and release oxygen. Both plants and animals depend on oxygen in the utilization of their food. Thus, plants provide a constant supply of this needed substance to our atmosphere.

In the revision the sentences are arranged more logically. Moreover, transitional elements — *thus* and *this needed substance* (referring to oxygen) — help link the last two sentences and clarify the main point of the paragraph.

Here is another paragraph to illustrate the importance of showing the reader a clear pathway of thought. Notice again the use of transitional words, along with the repetition of selected elements, to clarify the relationship between sentences:

According to sociobiological theory, the production of individual ova is costly relative to the production of sperm. *Therefore,* ova (or females) are the limiting factor in male reproductive success, and natural selection will favor those males who can compete effectively with other males for fertilizations. We can predict that males *will vary* greatly in fitness depending on their competitive abilities. *By contrast,* females *will vary* in their abilities to convert available resources into gametes and ultimately into viable offspring.

■ Make paragraphs work as integrated parts of the text.

Apart from the Abstract, most paragraphs are not isolated entities. Instead, they must mesh together smoothly as structural components of each section of your paper. This means that the beginning of one paragraph must "fit" with the end of the previous paragraph — you need to bridge the gap between the two passages gracefully. If you jump abruptly from one topic to the next, the text will seem choppy and disorganized and the reader will be confused. This problem often crops up in the Results and the Materials and Methods sections when you are presenting many different kinds of information.

You can use the same techniques for linking sentences *within* paragraphs for making transitions *between* paragraphs. These include the repetition of key words or ideas and the use of transitional "markers" (*furthermore, for example, a second point, by contrast, on the one hand . . . on the other hand,* and so on) to signal to the reader that you are either developing the same idea or moving on to a different one.

■ Vary your sentences.

Pay attention to the structure, length, and rhythm of your sentences. People "hear" writing as they read; if your prose is unvarying and one-dimensional you will not get your message across as effectively.

The following paragraph is dominated by short, choppy sentences:

Many doves showed "nest-calling" behavior. They assumed a position with the tail and body axis pointing slightly upwards. In this posture they flicked their wings. This behavior was seen in both sexes. It was especially common in males. I saw it performed both on and off the nest.

We can make this passage more readable (and therefore more interesting) by combining related sentences. This eliminates the distracting, choppy style and makes the paragraph more effective as a unit:

Many doves showed "nest-calling" behavior in which they assumed a position with the tail and body axis pointing slightly upwards and flicked their wings. This behavior, especially common in males, was performed both on and off the nest.

WRITING CLEAR, ACCURATE SENTENCES

■ Use words that say precisely what you mean.

Do not give in to the temptation to use a word that "sounds right" unless you are absolutely sure it is appropriate. Sentences that are otherwise perfectly effective can be ruined by a single word or phrase that is wrong for the context. Here are some examples from student papers for a mycology class:

INCORRECT — Kohlmeyer (1975) sees the pit plug of red algae as being *heavily* related to the plug of Ascomycetes.

CORRECT — Kohlmeyer (1975) sees . . . *closely* related . . .

INCORRECT — Many mycologists have spent years *researching* for data suggesting a red algal ancestry with higher fungi.

CORRECT — Many mycologists have spent years *searching.* . . .

INCORRECT — The evolutionary origin of the higher fungi has *harassed* scientists for many years.

CORRECT — The evolutionary origin . . . has *puzzled* scientists. . . .

Buy a dictionary and consult it often. Before you use a technical term, make sure you understand its meaning. If in doubt, look it up in the glossary of an introductory text. Or, check the reference section of your library for specialized dictionaries and scientific encyclopedias.

The following words are often used incorrectly in biological papers:

Affect: (as a verb) to influence or to produce an *effect.*
Effect: (as a noun) result; (as a verb) to bring about.

Nutrient concentration was the most important factor *affecting* population size.

Marking each ant on its thorax with enamel paint produced no apparent *effect* on its behavior.

We hope that further studies of these endangered species will *effect* a major change in the allocation of funds by the federal government.

Comprise: to contain or include. Do not use *comprise* when you should use *constitute* or *compose.*

INCORRECT	The vertebrate central nervous system is *comprised* of the brain and the spinal cord.
CORRECT	The vertebrate central nervous system *comprises* the brain and the spinal cord.
CORRECT	The vertebrate central nervous system is *composed* of the brain and the spinal cord.
CORRECT	The brain and the spinal cord *constitute* the vertebrate central nervous system.

Correlated: In scientific writing, use in conjunction with certain statistical tests (correlation analyses) that provide a measure of the strength of relationship between two variables. (See also p. 70.)

Interspecific: between or among two or more different species.

Intraspecific: within a single species.

> Leone (1960), studying four species of sandpipers at a Minnesota lake, found marked *interspecific* differences in food preferences.

> This plant shows little *intraspecific* variation in flower coloration; generally, the petals are pale yellow with a distinct orange spot at each tip.

Its/It's: Its is the possessive form of *it*. Do not confuse it with *it's*, a contraction of *it is* or *it has*.

> Each calf recognized *its* own mother.

> *It's* not clear whether rainfall or temperature is the more important factor.

Note, however, that contractions are generally avoided in formal biological writing.

Principal: (as an adjective) most important. Do not confuse it with *principle*, a noun, meaning a basic rule or truth.

> The *principal* finding of this study was that excessive drinking by rats did not cause significant increases in blood pressure.

> According to Heisenberg's Uncertainty *Principle*, we can never simultaneously determine the position and the momentum of a subatomic particle.

Random: In scientific writing, use to refer to a *particular* statistically defined pattern of heterogeneous values. (See also p. 70.)

Significant: In scientific writing, use to refer to *statistically significant* (or nonsignificant) results, after having conducted appropriate statistical tests. (See also p. 69.)

That/Which: Use *that* to introduce *restrictive* or defining elements — phrases or clauses that limit your meaning in some way.

> The rats *that had been fed a high calorie diet* were all dead by the end of the month.

Here the italicized portion restricts the meaning of *rats;* we are referring to *only* those specific rats.

Use *which* to introduce *nonrestrictive* or nondefining elements — word groups that do not limit your meaning but rather add additional information. Because this information is not vital to the integrity of the sentence, you can omit it without substantially changing the original meaning. The following sentence uses *which* instead of *that;* here the writer is speaking more generally, not calling attention to a *particular* group of rats.

> The rats, *which* were fed a high calorie diet, were all dead by the end of the month.

Misuse of *that* or *which* may make a sentence confusing:

> Plants, which grow along heavily traveled pathways, show many adaptations to trampling.

Here it sounds as if *all* plants grow along heavily traveled pathways. Actually the writer is referring only to a particular group of plants. The nonrestrictive clause needs to be replaced by a restrictive one limiting the meaning of *plants:*

> Plants that grow along heavily traveled pathways show many adaptations to trampling.

Use commas to set off nonrestrictive elements introduced by *which* — a *pair* of them if the element appears in the middle of the sentence.

> These data, which are consistent with those of other researchers, suggest several questions about the significance of wing positions in the thermoregulation of Arctic butterflies.

Do *not* use commas with restrictive elements and *that.*

Unique: one of a kind. A thing cannot be *most, very,* or *quite* unique; it is simply unique:

INCORRECT	Females of this species have a *very unique* horny projection from the dorsal part of the thorax.
CORRECT	Females of this species have a *unique . . .*

■ Avoid slang.

Slang is the informal vocabulary of a particular group of people. Some slang words eventually make their way into standard English; however, most soon become outdated and are replaced by others. Slang usually has no place in scientific writing even if you put it in quotation marks.

SLANG	The controversy over the evolutionary origin of the Ascomycetes and Basidiomycetes dates back almost a hundred years, but it has only recently moved to the cutting edge of research.

STANDARD	Recently, scientists have become more interested in the evolutionary origin of the Ascomycetes and Basidiomycetes, a topic that has been controversial for almost a hundred years.
SLANG	Barr's (1980) statement is merely a "cop out" because he refuses to acknowledge that there are major morphological differences between the two groups.
STANDARD	Barr (1980) fails to address this issue because he refuses to acknowledge that . . .

■ Revise misplaced modifiers.

Modifying words, phrases, or clauses should relate clearly to the words they are meant to describe. If they do not, the sentence may be confusing, even ludicrous.

FAULTY	After marking its hindwings with enamel paint, each damselfly was released within 1 m of the capture site.
REVISED	After marking its hindwings with enamel paint, I released each damselfly within 1 m of the capture site.

Unless these insects were particularly clever at marking themselves, we need to make the intended subject of this sentence (the writer) more evident. Using the active instead of the passive voice helps here. (See also pp. 145–47.)

FAULTY	After mating, the sperm are stored in a sac within the female damselfly's body.
REVISED	After mating, the female damselfly stores sperm in a sac within her body.

Who is actually mating, the sperm or the damselfly? The revised version corrects the ambiguity.

FAULTY	This behavior has only been reported in one other primate.
REVISED	This behavior has been reported in only one other primate.

Modifiers such as *only, even, almost,* and *nearly* should be placed next to the most appropriate word in your sentence, depending on your meaning. In the first sentence, the position of *only* before the verb suggests that behavior may have been observed in more than one primate but *reported* in just a single species. If you really want to say that this behavior has been observed *and* reported in just this one species, then *only* must modify *one* as in the revised version. In the following sentence, *only* does limit the meaning of the verb in accordance with the writer's meaning.

This behavior has only been observed casually, not reported formally.

■ Avoid vague use of *this, that, it,* and *which.*

Do not use these pronouns on their own to refer to whole ideas; you may lose the reader. To avoid confusion, use clear, specific wording.

VAGUE   We could not predict the number of adult males likely to visit each breeding site because male density in the surrounding forest varied greatly from day to day. This is typical of most field studies on this species.

What is "typical of most field studies"? Varying male densities, the difficulty of being unable to predict male numbers at the breeding site, or both? The revision eliminates this ambiguity:

SPECIFIC   We could not predict the number of adult males . . . from day to day. Varying male density is typical of most field studies. . . .

■ Make comparisons complete.

Add words if necessary to make comparisons or contrasts accurate and unambiguous.

AMBIGUOUS   Average body length in *Libellula pulchella* is longer than *Plathemis lydia.*

Here, you want to compare the body length of one species with the *body length* of another, as the revision clearly does:

UNAMBIGUOUS   Average body length in *Libellula pulchella* is longer than that in *Plathemis lydia.*

AMBIGUOUS   Bullfrogs were more abundant than any amphibian at Site A.

Because bullfrogs themselves are amphibians, you need to revise the wording to avoid confusion:

UNAMBIGUOUS   Bullfrogs were more abundant than any other amphibian at Site A.

■ Make each verb agree with its subject.

Do not lose sight of the subject in a sentence by focusing on modifying words, such as prepositional phrases, occurring *between* the subject and the verb.

The *size* of all territories *was* [not *were*] reduced at high population densities.

The *zygote* of the Ascomycetes *develops* [not *develop*] into ascospores.

Both sentences above have singular subjects and therefore need singular verbs.

Compound subjects are subjects with two or more separate parts that share the same verb. When these parts are connected by *and*, they need to be matched with plural verb forms.

The *color and shape* of the beak are [not *is an*] important taxonomic *features* [not *feature*].

When the parts of a compound subject are linked by *or* or *nor*, make the verb agree with the part that is closest to it. If one part of the subject is singular and the other plural, then put the plural part second and use a plural verb.

Under experimental conditions, *neither* the newly hatched offspring *nor* its older sibling *was tended* by the parents.

Here the parts of the compound subject are *offspring* and *sibling*; these are joined by *nor*. If the second part had been *siblings*, you would have needed a plural verb (*were tended*).

Do not confuse compound subjects with singular subjects that are linked to other nouns by prepositional phrases (such as *in addition to, along with, as well as*).

The dominant *male*, along with his subordinates, *protects* [not *protect*] the offspring when the troop is threatened by predators.

When you refer to a particular quantity of something as a single unit, treat it as singular.

Before each experiment, *10 ml* of distilled water *was* added to each vial.

Many scientific terms have a Latin heritage and are used in both the singular and plural form, depending on the context. Make sure the verb agrees with the subject.

The *larva* of the monarch butterfly *feeds* [not *feed*] on milkweed.

Here are the singular and plural forms of some words commonly used in biology.

SINGULAR	PLURAL
alga	algae
analysis	analyses
bacillus	bacilli
bacterium	bacteria
basis	bases

SINGULAR	PLURAL
criterion	criteria
datum (*rarely used*)	data
flagellum	flagella, flagellums
focus	foci, focuses
formula	formulae, formulas
fungus	fungi, funguses
genus	genera
hypothesis	hypotheses
index	indices (for numerical expressions), indexes (in books)
inoculum	inocula
larva	larvae
locus	loci
matrix	matrices
medium	media, mediums
mycelium	mycelia
nucleus	nuclei
ovum	ova
phenomenon	phenomena, phenomenons
phylum	phyla
protozoan	protozoa, protozoans
pupa	pupae
serum	sera, serums
species	species
spectrum	spectra
stimulus	stimuli
stratum	strata
symposium	symposia, symposiums
taxon	taxa
testis	testes
villus	villi

Note that the word *species* can be either singular or plural: one species of cat, three species of toads.

■ Put related elements in parallel form.

When you link two or more words, phrases, or clauses in a sentence, put them in the same grammatical form. Such parallelism makes your writing easier to read and emphasizes the relationship between items or ideas.

FAULTY	These two species differ in color, wingspan, and where they typically occur.
PARALLEL	These two species differ in color, wingspan, and habitat.

FAULTY	Both populations of plants had high mortality rates at sites that were dry, windy, and where there were frequent disturbances.
PARALLEL	Both populations of plants had high mortality rates at sites that were dry, windy, and frequently disturbed.

FAULTY	Male bluntnose minnows promote the survival of their offspring by agitation of the water over the eggs and keeping the nest free from sediment.
PARALLEL	Male bluntnose minnows promote the survival of their offspring by agitating the water over the eggs and keeping the nest free from sediment.

In the first example, the clause, *where they typically occur*, has been replaced by a single noun, *habitat*, for parallelism with the two preceding nouns (*color, wingspan*). Similarly, in the second example, parallelism has been achieved by putting *where there were frequent disturbances* into adjective form, *disturbed*, to agree with *dry* and *windy*. In the third example, the noun *agitation* has been changed to a present participle (verb form ending in -*ing*), making two parallel phrases, "*agitating . . .*" and "*keeping. . . .*"

■ Write in a direct, straightforward manner; avoid jargon.

Scientific writing has the reputation of being dry, monotonous, and hard to understand. Consider the following passage from a published paper.

> The data of this study suggest that such a handling procedure not only effects a diminution of emotional behaviour as indexed by decreases in the duration of the immobility reaction, but also as indexed by other measures of fear (freezing and mobility in the open field). To the extent that the handled and nonhandled groups differed in the predicted direction with respect to these indices of fear, distress vocalizations in the open field were shown to be significantly less frequent in the relatively more fearful (non-handled) group. (Ginsberg, Braud, and Taylor 1974, p 748)

A paper may contain exciting results or brilliant insights, but still be tedious to read because of the author's unengaging prose. Whether it is a manuscript submitted for publication or a laboratory report for a biology class, an impenetrable style may keep it from getting the attention it deserves.

Frequently, the problem is the use of *jargon* instead of simpler, more straightforward writing. Broadly, jargon is the technical language of some specialized group, such as biologists. More specifically (and negatively), jargon is long-winded, confusing, and obscure language. Writers of jargon use esoteric terms unfamiliar to most of their readers. They rely heavily on long sentences, big words, a pompous tone, and ponderous constructions in the passive voice.

JARGON	One hour prior to the initiation of the experiment, each avian subject was transported by the experimenter to the observation cage. The subject was presented with various edible materials, and ingestion preferences were investigated utilizing the method developed by Wilbur (1965). When data collection was finalized, the subject was transferred back to the holding cage.
REVISED	One hour before the experiment, I put each bird in the observation cage, where feeding preferences were studied using Wilbur's (1965) method. The bird was then returned to the holding cage.

The revised passage conveys the same information as the jargon-ridden one, but more simply, directly, and concisely.

Do not assume that to sound like a biologist you must write dry, stilted, and boring prose, or that complex ideas must be couched in equally complex, convoluted sentences. Biology instructors, along with editors and readers of biological journals, prefer clear, straightforward writing — simple but effective prose that quickly illuminates the author's results and ideas.

AVOIDING WORDINESS

First drafts are usually labored and wordy because you have been focusing on just writing down your ideas. As you revise, examine each sentence carefully. Could you say the same thing more succinctly without jeopardizing the content? Lifting just one excess word from a sentence can enliven its rhythm and intensify its meaning, making your prose carry more weight.

■ Omit unneeded words; shorten wordy phrases.

WORDY	On two occasions, I succeeded in observing a mating pair for the entire duration of copulation.
CONCISE	I observed two pairs for the duration of copulation.
WORDY	There now is a method, which was developed by Jones (1973), for analyzing the growth of rotifer populations.
CONCISE	Jones (1973) developed a method to analyze the growth of rotifer populations.

WORDY	The sample size was not quite sufficiently large enough.
CONCISE	The sample size was not large enough.
MORE CONCISE	The sample size was too small.

WORDY	The root cap serves to protect the cells of the root meristem as the root is growing through the soil.
CONCISE	The root cap protects the cells of the root meristem as the root grows through the soil.

WORDY	The eggs were blue in color, and they were covered with a large number of black spots.
CONCISE	The eggs were blue with many black spots.

Common modifiers such as *very, quite,* and *rather* can often be cut from sentences. You may use such words routinely without asking yourself if they are essential to your meaning.

FAULTY	The data in Table 1 are *very consistent* with Leshchva's (1966) model.
REVISED	The data in Table 1 are *consistent* with . . .

FAULTY	Males guarding eggs are *quite aggressive* toward juveniles and females.
REVISED	Males guarding eggs are *aggressive* toward . . .

In summary, biological writing is plagued by countless wordy phrases, often placed at the beginning of a sentence, and by "empty" words and phrases that add little to the author's meaning. Some of these are listed below. (See also Day 1994.) You will probably think of many others to add to the list.

WORDY	**CONCISE**
a second point is that	second, secondly
more often than not	usually
it is apparent that	apparently
at the present time	now
in previous years	previously
owing to the fact that	because
because of the fact that	because
in light of the fact that	because
it may be that	perhaps
these observations would seem to suggest	these observations suggest
one of the problems	one problem

WORDY	CONCISE
in only a very small number of cases	occasionally, rarely
in the possible event that	if
An additional piece of evidence that helps to support this hypothesis	Further evidence supporting this hypothesis
In spite of the fact that our knowledge at this point is far from complete	Although our present knowledge is incomplete
It is also worth pointing out that	*omit it*
Before concluding, another point is that	*omit it*
It is interesting to note that	*omit it*

■ Avoid repetition.

Some sentences or paragraphs are wordy because the writer has included the same information twice.

WORDY	In Kohmoto's study in 1977, she failed to account for temperature fluctuations (Kohmoto 1977).
CONCISE	Kohmoto (1977) failed to account for temperature fluctuations.

Because the author's name and the publication date are given in the literature citation (Kohmoto 1977), you need not give the same information at the beginning of the sentence. (See also Chapter 6.)

In the next example, the second sentence can be omitted because it merely repeats part of the first sentence using different wording.

> Male fathead minnows who were tending eggs spent the majority of their time rubbing the egg batches with their dorsal pads and preventing other fish from eating the eggs. They did not devote much time to other activities, but instead were chiefly occupied with behavior directed towards the eggs.

■ Use the passive voice sparingly.

In the passive voice, the subject of the sentence *receives* the action, whereas in the active voice it *does* the action.

PASSIVE	Nearly half the seedlings *were eaten* by woodchucks.
ACTIVE	Woodchucks *ate* nearly half the seedlings.

Biological writing leans heavily on the passive voice even though the active voice is more direct, concise, and effective. Why this emphasis on the passive? Compare the following two sentences:

PASSIVE	Skin extract solution *was presented* to the fish through a plastic tube.
ACTIVE	*I presented* skin extract solution to the fish through a plastic tube.

Neither sentence is incorrect; however, the sentence in the passive voice shifts the reader's attention away from the writer and more appropriately to the materials he or she has been testing, observing, collecting, or measuring. This is why the passive voice is particularly common in the Materials and Methods section of biological papers. Shunning the passive altogether may result in prose that is "I-heavy" and monotonous because of too many first-person references. Prudent use of passive constructions gives variety to the text, and in a few situations such constructions may simply be more convenient.

Thus the passive voice has legitimate uses. Unfortunately, however, many beginning writers rely heavily on the passive because they think it makes their prose more formal, more important, more "fitting" for a scientist. When used habitually, carelessly, or unintentionally, the passive voice results in a wordy and cumbersome style. Overuse of the passive may be one reason why scientific writing has the reputation of being dry, pompous, and boring. Usually the passive voice can easily be converted to the active voice, making sentences shorter and more forceful without any loss of meaning.

PASSIVE	Territory size was found to vary with population density.
ACTIVE	Territory size varied with population density.
PASSIVE	From field observations, it was shown that virtually all tagged individuals remained in their original home ranges.
ACTIVE	Field observations showed that virtually all tagged individuals remained in their original home ranges.
PASSIVE	Nest destruction was caused primarily by raccoons, particularly late in the incubation period, when greater access to nests was afforded to them by lowered water levels.
ACTIVE	Raccoons caused most nest destruction, particularly late in the incubation period when lowered water levels afforded them greater access to nests.

The passage below, from the Methods section of a paper by Burger (1974, p 524), illustrates an effective mix of active and passive voice.

I used several marking techniques on nests, adults, and juveniles. In 1969, I marked nests with red, blue, and white plastic markers tied to cattails. . . . Markers placed on nests were subsequently covered with fresh nest material. Adults were captured with a nest trap . . . and marked with coloured plastic wing tags (Saflag). I marked pairs who were close to the blind by pulling a string attached to a cup of red dye suspended over the nest.

Notice that the author of this passage uses several first-person references (*I* used . . . ; *I* marked . . .). Today many biologists are writing "I" instead of impersonal and cumbersome language such as "this investigator" or "the author." This use of the first person makes their prose more direct and concise. It may also reflect a growing realization by biologists that there *is* an "I" in science — that scientific research is inevitably influenced by the personal background, interests, motives, and biases of each researcher.

You will still find the passive voice in most of the scientific literature you read. As you write your own papers, you may feel the need to use the passive voice. There are no firm guidelines about this in biological writing, and different journals and scientific fields vary in their use of the passive. However, the active voice generally does a better job. Use passive constructions deliberately and sparingly. For every sentence you put in the passive, ask yourself if you could express it more exactly and concisely in the active voice.

VERB TENSE

Scientific ethics have given rise to the convention of using the *past* tense when reporting your own present findings and the *present* tense when discussing the published work of others. This is because new data are not yet considered established knowledge, whereas the findings of previous studies are treated as part of an existing theoretical framework. Therefore, in a research paper you will need to use both the past and present tenses. Most of the Abstract, Materials and Methods, and Results sections will be in the past tense because you are describing your own work. Much of the Introduction and Discussion will be in the present tense because they include frequent references to published studies. Look at the following examples.

The reproductive success of yellow-bellied marmots (*Marmota flaviventris*) is strongly influenced by the availability of food and burrows (Andersen, Armitage, and Hoffmann 1976) [Introduction].

After inoculation, plants were kept in a high-humidity environment for 100 h [Materials and Methods].

Limpets occurred at all sampling sites [Results].

Diaptomus minutus was dominant in the zooplankton of Clinton Lake during both years of the study. This species *is* common on many other acidic lakes in the region (Lura and Lura 1985) [Discussion].

When you refer to an author directly, however, you may use the past tense:

Bauman (1959) found that this bacterium is highly sensitive to pH.

When you refer directly to a table, a figure, or a statistical test in your own paper, it is acceptable to use the present tense.

Table 3 shows that polychaetes were most abundant at depths of 10–16 m.

See the sample research paper at the end of Chapter 4 for other examples of correct verb tense.

PUNCTUATION

Used correctly, punctuation marks help make your writing clear and understandable. Used incorrectly, they may distract, annoy, or confuse the reader. Buy a writer's guide with a detailed section on punctuation, and consult it frequently; several good handbooks are listed at the end of this book. The guidelines below address some of the most common punctuation problems in biology writing.

Comma

1. Use a comma to separate introductory material from the rest of the sentence.

Although egg cases were reared using Hendrickson's (1977) method, none of the eggs hatched.

During the 1987 breeding season, bullfrogs were sexually active from early June to late July.

Nevertheless, these data suggest that mate selection in this species is based primarily on female choice.

A comma may be omitted after a very *short* introductory element that merges smoothly and unambiguously with the rest of the sentence.

Thus the data in Table 1 are consistent with those of Tables 2 and 3.

2. Use a *pair* of commas to set off insertions or elements that interrupt the flow of a sentence.

The situation is different, however, on isolated islands with lower species diversity.

This explanation, first proposed by Hess (1967), is still widely accepted.

3. Use a comma to separate all items in a series, including the last two.

Unlike mosses, ferns possess true roots, stems, and leaves.

4. Use a comma to separate independent clauses linked by a coordinating conjunction (*and, but, or, for, nor, so, yet*). (An independent clause is a word group containing a subject and a verb and able to function as a complete sentence.)

Many studies have been made of feeding preferences in spiders, but few have been done under natural conditions.

Do *not* use a comma on its own to join two independent clauses. Such an error is called a *comma splice*.

COMMA SPLICE   Horsetails usually grow in moist habitats, some occur along dry roadsides and railway embankments.

A comma splice also results when you join two independent clauses with a comma followed by a conjunctive adverb (*however, moreover, furthermore, therefore, nevertheless*).

COMMA SPLICE   Horsetails usually grow in moist habitats, however some occur along dry roadsides and railway embankments.

There are several ways to correct a comma splice. For example, you can use a semicolon (see p. 150) with or without a conjunctive adverb:

REVISION 1   Horsetails usually grow in moist habitats; however, some occur along dry roadsides and railway embankments.

REVISION 2   Horsetails usually grow in moist habitats; some occur along dry roadsides and railway embankments.

A comma splice can also be corrected by adding a coordinating conjunction after the comma:

REVISION 3   Horsetails usually grow in moist habitats, but some occur along dry roadsides and railway embankments.

Alternatively, you can make one of the clauses a dependent clause, and separate the two clauses by a comma. (A dependent clause lacks either a subject or a verb and cannot stand alone as a complete sentence.)

REVISION 4   Although horsetails usually grow in moist habitats, some occur along dry roadsides and railway embankments.

Finally, you can make two sentences separated by a period:

REVISION 5   Horsetails usually grow in moist habitats. Some occur along dry roadsides and railway embankments.

Semicolon

1. Use a semicolon to join two independent clauses *not* linked by a coordinating conjunction (*and, but, or, for, nor, so, yet*).

> Bacteria reproduce very rapidly; many species can divide once every 20 minutes under favorable conditions.

This statement could also have been written as two complete sentences separated by a period, but the semicolon suggests a closer relationship between the two ideas.

2. Use a semicolon to connect two independent clauses joined by a conjunctive adverb (*however, moreover, nevertheless, furthermore*).

> These data are consistent with those of Allen (1980); moreover, they suggest that pH may be a more important influence than previously believed.

3. Use a semicolon to clarify the meaning of a series of items containing internal commas.

> Films were made of courtship and mating behavior; aggressive interactions between males, including chasing, butting, and biting; and parental behavior by females, particularly fanning and egg retrieving.

Colon

1. Use a colon to introduce items in a series.

> Birds are distinguished by the following features: a four-chambered heart, feathers, light bones, and air sacs.

However, do *not* use a colon before a series unless the colon follows an independent clause.

> **INCORRECT**   Dandelions are in the same family as: daisies, chrysanthemums, sunflowers, and hawkweeds.

> **INCORRECT**   The most common insects at Site A were: mayflies, dragonflies, damselflies, and butterflies.

In both these examples, the colon should be deleted.

2. Use a colon between two independent clauses when the second clause explains or clarifies the first one.

> The Cranston Lake study site was most suitable for incubation studies: nest sites were abundant, and brooding birds were rarely disturbed by road traffic.

3. Use a colon to formally introduce a direct quotation. (See pp. 98–99.)

Dash

The dash is used to separate material abruptly from the rest of the sentence for clarity, emphasis, or explanation. (To type a dash, use two hyphens, one after the other, with no space on either side: --. Word processing programs will then convert these hyphens to dashes.)

> Insects are distinguished by three body regions — head, thorax, and abdomen — and three pairs of legs.

Many writers use the dash when a comma, indicating just a slight pause, would be less distracting.

DISTRACTING	Eggs with minor cracks and an intact inner membrane hatched normally — but severely cracked eggs failed to develop and soon rotted.
REVISED	Eggs with minor cracks and an intact inner membrane hatched normally, but severely cracked eggs failed to develop and soon rotted.

When overused, dashes make your writing seem informal and careless, suggesting that you do not know how to use other types of punctuation. Use dashes sparingly, if at all, in biological writing.

Parentheses

1. Use parentheses to insert explanatory or supplemental information into a sentence.

> The nests at Leland Pond were unusually large (at least 30 cm in diameter), but only one contained any eggs.

> Juvenile monkeys raised in groups of three gained weight more rapidly than those raised in isolation (Fig. 2).

Like dashes, parentheses should be used sparingly because they interrupt the flow of your writing. Commas often work just as well as parentheses and are less distracting.

2. Use parentheses with a series of items introduced by letters or numbers, especially when you are listing many items or several long items.

> Dragonflies are good subjects for studies of territorial behavior for these reasons: (1) many are easily identified and widely distributed, (2) most can be observed easily under natural conditions, and (3) much is known about the anatomy and life history of many species.

Quotation Marks, Brackets, Ellipsis

See pages 98–99 for uses of these punctuation marks to present quoted material.

CHAPTER 8

Preparing the Final Draft

MECHANICS AND TECHNICALITIES

- ### Use the scientific names of organisms.

The common names of plants and animals have arisen through long traditions of folklore and popular usage. Many reflect some aspect of an organism's appearance, natural history, or relevance to human life: lady's slipper, hedgehog, bedbug, horsehair worm, liverwort. However, common names can also be misleading. Ladybugs, strictly speaking, are not bugs (Hemiptera) but beetles (Coleoptera), and club mosses (Lycopsida) are not true mosses (Musci). Because common names are not universally agreed upon, a single species may be called by different names in different localities. Also, the same common name may be used for two or more taxonomically unrelated organisms.

However, each species of organism has just one scientific name. It consists of two parts: first, the word denoting the *genus;* and second, the specific epithet (or *species*). Both typically are Latin words or Latinized forms. For example, the scientific name of red maple is *Acer rubrum; Acer* is the genus (to which many different species of maples belong). *Acer rubrum* is the particular species, red maple.

The scientist who first publishes the scientific name of a newly recognized species is regarded as its "author"; his or her last name is placed after the generic and specific names: *Pimephales promelas* Rafinesque. An author's name is often abbreviated: R. for Rafinesque, L. for Linnaeus, Fab. for Fabricius. (However, sources may vary with respect to the abbreviations used for the same author's name.) Sometimes an author's name is put in parentheses, indicating that the species has been put into a different genus from the original. In botany and microbiology, but not zoology, the name of the person responsible for this new classification is also added at the end. For example, in the case of *Kickxia spuria* (L.) Dum., the plant was originally called *Linaria spuria* by Linnaeus but was later placed in a different genus, *Kickxia*, by Dumortier (Dum.).

Thus, the scientific names of organisms are determined according to a system of universally accepted rules and conventions. Biologists rely heavily on scientific names; they use common names less frequently. (In fact, most species known to science lack common names.) It is acceptable to refer to organisms by common names as long as you have first given full scientific names along with other taxonomic information, if necessary. Here are some general guidelines for using scientific and common names:

1. The scientific names of species are always italicized (or underlined if you are writing in longhand or using a typewriter). The genus name is capitalized; the specific epithet is not: *Plumularia setacea*. Subspecies or other names below the level of species are italicized but not capitalized: *Plectrophenax nivalis subnivalis*. Author names or their abbreviations are capitalized but not italicized: *Haematosiphon inodorus* (Duges).

2. Give the full scientific name the first time a particular species is referred to in the text. In later references to this species you may abbreviate the genus by its first letter (still capitalized and italicized). The bacterium *Spirillum volutans* thus becomes *S. volutans*. Confusion may arise, however, if you are discussing two or more organisms whose generic names begin with the same letter — for example, *Spirillum volutans* and *Streptococcus salivarius*. In that case, it is safer to spell out the generic names.

Also, spell out the generic name at the start of a sentence, or else reword the sentence.

INCORRECT	*S. vicinum* was the most common dragonfly at Site A.
CORRECT	*Sympetrum vicinum* was the most common dragonfly at Site A.
CORRECT	The most common dragonfly at Site A was *S. vicinum*.

3. Generally, the author of a species is specified only the first time the species is mentioned in the paper. Often this information is included in the title, as well.

4. Do not put an article (*the, a, an*) immediately before the scientific name of a species.

INCORRECT	The most common lichen at both sampling sites was the *Lecidea atrata*.
CORRECT	The most common lichen . . . was *Lecidea atrata*.

5. Do not pluralize scientific names.

INCORRECT	Digger bees (*Centris pallidas*) were also observed at the study site.
CORRECT	Digger bees (*Centris pallida*) were . . .

6. Except in keys (identification guides) or other taxonomic writings, the specific epithet of an organism must always be preceded by the generic name or its abbreviation.

INCORRECT	The most common species at Hamilton Creek were *Achnanthes minutissima* and *Meridion circulare*. *Circulare* was also the most common alga at Woods Creek later in the sampling period.
CORRECT	The most common species [was] . . . *Meridion circulare*. *Meridion circulare* was also . . .

The names of genera, however, may be used alone if you are referring collectively to the species in a particular genus.

Insecticides have been used in an attempt to eradicate *Anopheles* mosquitoes and thus control the spread of malaria.

Some species of *Sargassum* grow in dense mats on the surface of the ocean.

7. Taxonomic groups, or taxa, above the level of genus (family, order, class, phylum, division, and so on) are capitalized but not italicized or underlined.

The Chilopoda (centipedes) and the Diplopoda (millipedes) are two of six classes in the subphylum Mandibulata.

8. Biologists frequently drop or modify the endings of taxa to make common names for organisms — for example, chironomids from Chironomidae; lycopsids from Lycopsida; dipterans from Diptera; cephalopods from Cephalopoda. Such words, unlike the formal names for groups, are not capitalized.

9. Similarly, other common names of organisms are not usually capitalized except in accordance with specific taxonomic guidelines for certain groups. For example, the common names of American birds have been standardized by the American Ornithologists' Union and *are* capitalized: American Robin, Chipping Sparrow, Barn Swallow. Shortened forms of these proper common names are not capitalized: robin, sparrow, swallow. If a common name contains a word derived from the name for a particular person or place, then that word is generally capitalized even if the rest of the common name is not: English ivy, Queen Anne's lace.

10. Biologists use conventional abbreviations to refer to one or more undesignated species of a particular genus:

Most nematodes collected from Site 4 were affected by a fungal parasite (*Myzocytium* sp.) [one species].

Francini (1970) found that in certain butterflies (*Colias* spp.) . . . [more than one species].

■ Record time according to a 24-hour system.

Biologists do not report times as AM or PM. Instead, they use a 24-hour time system, numbering the hours of the day consecutively starting at midnight (0000 hours). For example:

7:30 AM	=	0730
12:00 noon	=	1200
9:45 PM	=	2145
11:53 PM	=	2353

Biological writing also uses a convenient shorthand to specify *photoperiod* (the number of hours of light in a 24-hour period). For example, instead of writing, "Hamsters were exposed to a photoperiod of 12 h," a writer could say, "Hamsters were reared under conditions of 12L:12D" (12 hours of light followed by 12 hours of darkness). Similarly, 14L:10D means 14 hours of light and 10 hours of darkness.

■ Use symbols and abbreviations commonly used in biology.

Following are some symbols and abbreviations commonly used in biology. Note that many are not followed by a period (unless they appear at the end of a sentence). There is some variation from one scientific journal to the next. The Council of Biology Editors (CBE) style manual (see p. 194) contains extensive lists of recommended symbols and abbreviations for biologists.

TERM/UNIT OF MEASUREMENT	SYMBOL/ ABBREVIATION
ångström	Å
approximately	ca. *or* c. *or* ≈
calorie	cal
centimeter	cm
cubic centimeter	cm^3
cubic meter	m^3
cubic millimeter	mm^3
day	d
degree Celsius	°C
degree Fahrenheit	°F
diameter	diam
degrees of freedom	df, v

Term/Unit of Measurement	Symbol/Abbreviation
east	E
et alia (Latin: and others)	et al.
et cetera (Latin: and so forth)	etc.
exempli gratia (Latin: for example)	e.g.
female	♀
figure, figures	Fig., Figs.
foot-candle	fc *or* ft-c
gram	g
greater than	>
hectare	ha
height	ht
hour	h
id est (Latin: that is)	i.e.
joule	J
kelvin	K
kilocalorie	Kcal
kilogram	kg
kilometer	km
latitude	lat.
less than	<
liter	l *or* L
logarithm (base 10)	log
logarithm (base *e*)	ln
longitude	long.
male	♂
maximum	max
mean (sample)	$\bar{x}, \bar{X}, \bar{y}, or \bar{Y}$
meter, metre	m
microgram	μg
microliter	μL
micrometer	μm
milliliter	mL
millimeter	mm
minimum	min
minute (time)	min
molar (concentration)	mol L^{-1}
mole	mol
month	mo
nanometer	nm
north	N
not (statistically) significant	NS *or* ns
number	nr
number (sample size)	*n*
parts per million	ppm

Term/Unit of Measurement	Symbol/ Abbreviation
percent	%
plus or minus	$\pm$
probability	*P*
second (time)	s
south	S
species (singular)	sp.
species (plural)	spp.
square centimeter	cm^2
square meter	m^2
square millimeter	mm^2
standard deviation	*s or* SD
standard error	$s_{\bar{x}}$ *or* SE
standard temperature and pressure	STP
versus	vs.
volume	vol
volt	V
watt	W
week	wk
weight	wt
west	W
year	y

MANUSCRIPT FORMAT

After making the final revisions in a piece of writing, you may feel that all the work is over. It is — almost. You still need to produce a neat, clean manuscript. A paper with page numbers missing, margins askew, and the type barely readable will not show off your prose to good advantage, and it may suggest that little care was taken in more substantive matters. At the very least, a carelessly prepared manuscript creates a bad first impression, and in some instances it may not be read at all. Now that computer technology has so simplified the basic, mechanical process of manuscript preparation, there simply is no excuse for a sloppily prepared document.

Biologists submitting work for publication generally must adhere to specific format guidelines when preparing the final manuscript. In the case of a research paper, for example, there may be editorial requirements about the number of manuscript copies to be submitted; the paper quality and typeface; the margins and spacing; the contents of the title page; the methods used for documentation; the form in which figures, tables, and other visual materials are to be prepared; and so on. Prospective authors who fail to follow the stylistic and format conventions of the journal for which they are writing virtually guarantee that their manuscript will receive little serious attention.

For academic assignments, too, a neatly prepared manuscript suggests that you are serious about your work and have devoted care and thought to all stages of its preparation. Check if your instructor has his or her own format specifications. Otherwise, the guidelines below will help you produce a professional-looking manuscript.

Paper, Margins, and Spacing

Use 8½-by-11-inch white paper and a standard typeface (not script type or all capitals). A font size of 11 or 12 points is typical. Print on only one side of the paper. Leave margins of 1 to 1½ inches on all sides of the page, and do not justify the right margin. Indent each new paragraph five spaces.

Biologists submitting manuscripts for publication are usually asked by journal editors to double-space the entire manuscript, including the Abstract and the Literature Cited section. This format is preferable, too, for academic assignments. Another option is to use single-spacing within each entry in the Literature Cited section, with double-spacing between entries. The Abstract may also be single-spaced. Check with your instructor to see if he or she has spacing preferences.

Journals vary with respect to whether indentation is used in the Literature Cited section. In academic papers, a common practice is to place the first line of each entry flush with the left margin, and indent successive lines of the entry three to five spaces.

Title Page

An acceptable format for a student paper calls for the following information on the title page: title of the paper, your name, course information, and the date. Center the title and place it about one-third down the page; center and double-space all successive lines under the title. It is not necessary to underline the title, to put it in quotation marks, or to follow it with a period. Capitalize the first letter of all important words on the title page — that is, words other than articles (*a, an, the*), coordinating conjunctions (*and, but, or, for, nor, so, yet*), and prepositions (*in, on, to,* and the like).

Pagination

Number pages consecutively beginning with the title page (which does not actually carry a number, but is still counted). Use only arabic numerals (1, 2, 3, . . .), and put the number in the upper right-hand corner of each page. Writers submitting a research paper for publication often begin each major section of the paper (Abstract, Introduction, Materials and Methods, and so on) on a separate page. This is not necessary for a student paper, although you may find that separating the sections makes a long paper easier to follow.

Headings

Headings are useful organizational tools. They can designate major portions of a work (for example, the Introduction or the Discussion of a research paper), and they can be used in a review paper to divide a long or complex text into smaller, more coherent, and more manageable sections. The following guidelines will help you use headings correctly.

1. Use headings and subheadings sparingly. Student papers generally need no more than one or two levels of headings. Overuse of headings results in a fragmented text, with many small sections. As a result, readers may have difficulty following your organizational logic and seeing how the various sections fit together.

2. Make headings informative and concise. Headings function like signposts, helping to orient readers. They should not, themselves, be cumbersome and tedious to read.

3. Use the same format for headings on the same level of organization. For example, you may decide to center and capitalize all first-order headings, and to underline second-order headings at the left-hand margin:

```
                    RESULTS

Outcrossing rate

As shown in Figure 1, the proportion of spores

that formed appressoria at the end of germ

tubes . . .
```

Whatever you decide, however, be consistent, or else you will confuse your readers.

4. Use parallel grammatical form for headings on the same level of organization. (A common practice is to use all nouns or noun phrases, or all verb forms ending in *-ing*.) Whatever grammatical form you choose, be consistent. For example, suppose that your first two headings in the Results section of a lab report were *Mode of infection* and *Temporal variation of infection*. Subsequent headings should also be noun phrases: not *Measuring spatial variations of infections,* but *Spatial variations of infections, Consequences of infection,* and so on.

Figures and Tables

Biological journals usually require authors to put each table or figure on a separate page. The title of a table is typed on the same page as the table; however, figure legends are typed together on a separate page. Tables, figures, and figure legends are then placed at the end of the Literature Cited section in that order. You may wish to follow this procedure for academic assignments, unless your instructor prefers that tables and figures be incorporated with the text. If so, put each table or figure after the page on which it is first mentioned and number all pages. In this case, each figure should be accompanied by its legend, typed on either the same or a separate sheet.

Still another way to deal with figures and tables is to insert them directly onto pages containing text, again putting each figure or table as close as possible to the point at which it is first mentioned. This will make your paper resemble a published work. You can do it on a computer: most good word processors can import graphics into a text file, allowing you to position a figure or table on a particular page, with writing either above or below it.

Assembling the Manuscript

Never submit a loose pile of pages: use a staple or a paper clip to fasten everything together. A special folder is not necessary (although it's a nice touch). In case your instructor misplaces the paper, keep a copy of the original on disk or make a photocopy of the original.

WRITING AN ACKNOWLEDGMENTS SECTION

In published biological papers, a short Acknowledgments section usually comes between the Discussion and the Literature Cited sections. Here you express thanks to people who assisted you with the research itself or with the preparation of the paper. Scientific ethics dictate that you first obtain the permission of all those you intend to acknowledge, preferably by showing them a copy of that section of the paper. This is because it is conceivable (although probably unlikely) that some people who offered you help may not wish to be associated in print with the paper, or because they may object to the wording you used to refer to them. See the sample papers in Chapters 4 and 5 for examples of Acknowledgments.

PROOFREADING

Typographical errors, misspelled words, missing commas or periods, irregular spacing, and other minor errors distract the reader and undermine your authority as writer. The aim of proofreading is to eliminate such mistakes from the final draft of the manuscript.

It is not easy to proofread well. By the time you have worked through several drafts of the paper you are no longer as observant as you might be. You will tend to read what you *think* is there, not what is actually on the page. Therefore you must force yourself to look at every word, space, line, punctuation mark, and number, including all material in figures and tables. Proofread the whole manuscript more than once over a period of several hours or days; you may miss some errors the first time but notice them during the second or third proofreading, when you have achieved more "distance" from the material. Ask a friend to look over the paper as well.

Finally, learn to use word processing features that speed up revising and proofreading (see pp. 129–30).

CHAPTER 9

Using Writing to Prepare for Examinations

That writing facilitates learning is widely recognized. Conversely, if you do not fully understand something, you will probably have difficulty writing about it. Often the process of finding the right words can help correct flaws in your reasoning and clarify your thinking. Even the *mechanical* act of writing, whether by hand or on the computer, gets you more involved with the material. Writing *as* you read or think reinforces your memory of the facts and can reveal connections between related ideas.

GETTING THE MOST OUT OF TEXTBOOKS

Following are several ways to make reading your course texts an *active,* not a *passive,* experience, and thus better prepare you for labs and exams.

■ Use highlighting sparingly.

You may be used to emphasizing important parts of a text by underlining them or using a highlighter pen. This is a useful reading strategy but only if used thoughtfully. Avoid highlighting line for line *as* you are reading; it doesn't allow you to make meaningful distinctions about the relative importance of successive sentences. It is better to *first* read a passage all the way through; then go over it again and highlight *selective* points. The reading will take longer, but you will learn more this way.

■ Annotate your textbooks.

Adding your own annotations — handwritten questions, notes, and other comments in the margins — is another way to read interactively. Make difficult passages comprehensible by summing them up in your own words. Insert questions next to material you don't understand. If certain points relate to lectures or lab exercises, make notes to that effect so that you can integrate various components of the course. Used selectively with highlighting, annotation can individualize your texts and make information easier to access and retain.

■ Read chapter summaries and respond to review questions.

These study aids reflect the author's expert insights about the most important material. If your text does not have a chapter summary, write your own. If a summary is provided, try composing your own version anyway. Write out the answers to end-of-chapter questions, even ones you think you understand; do not simply answer them in your head.

TAKING GOOD LECTURE NOTES

Over the years, you have probably developed your own strategies for note taking. By now you know that good class notes can be invaluable for self-teaching. The following suggestions will help you use writing more effectively to study biology and many other subjects.

■ Be selective.

During lectures, do not feel you have to record every word. If you write continuously, you will not be giving full attention to what the speaker is saying. Your notes may be voluminous, but they may not make sense because you won't be thinking about what you are writing. Taking notes nonstop also inhibits your ability to ask questions about what you do not understand and to contribute to class discussions.

Good lecture notes are a selective record. They reflect the writer's attempts to summarize and synthesize the most important points, to reproduce the logic and continuity of the speaker's discussion, and to make links to material from previous class meetings. Many of your notes may be abbreviated versions of the lecturer's own sentences, but others should be in your own words, particularly when you need to bridge the gap between one topic and another.

■ Revise your notes as soon as possible.

You probably use various shortcuts to take notes quickly, such as writing in sentence fragments and abbreviating words. Your writing may make sense at the time but be hard to decipher later. For this reason, review each day's notes while the material is still fresh in your mind. Write out any shortened sentences or words that may confuse you later. Clarify difficult concepts by adding comments in the margins. Check that diagrams are accurately labeled and integrated with the rest of your notes. Number related items and use underlining, asterisks, or other labels to call attention to the most important ideas. Reworking your notes after each lecture may seem time-consuming, but it will reward you later when you are studying for examinations. Moreover, it reinforces your understanding of the material.

■ Keep up with assigned readings.

Instructors write their lectures assuming students have acquired certain background knowledge by reading the textbook or other course materials. If you come to class with that knowledge, you are more likely to understand the lecture and to take coherent notes. Moreover, you will be less likely to waste time and energy writing down information covered adequately elsewhere.

PREPARING FOR LABORATORY EXAMS

Introductory and broad survey courses in biology often have laboratory exams, or practicals, as well as lecture examinations. Preparing for a lab exam may be difficult if you have not taken good notes during the laboratory periods. Much of the advice listed for lecture notes also applies here. Remember the benefit of advance preparation: read both your text and your lab manual, and understand how each lab's exercise fits into the framework of the course as a whole. Record observations and questions as you work; do not trust your memory.

When preparing drawings, use your text or other references to help you identify structures, but draw what you *see;* do not simply copy a stylized textbook illustration. Suppose for instance, you wish to draw a cross-section of a *Smilax* root from a microscope slide (see Fig. 1 and sample student drawing based on it). You need not be an artist to render an accurate, detailed, and comprehensible illustration. Use pencil for your initial sketch, and allow enough room to include sufficient detail. Label your diagram carefully and always print legibly; be sure to identify details using the correct terminology. Check that all lines pointing to particular structures are

Figure 1. Photomicrograph of *Smilax*
root in cross-section (100 X)

POOR

IMPROVED

epidermis
(cells have thick walls)

exodermis
(cells have very thick
walls except at base
– stained pink)

large parenchyma
cells of cortex
(contain insoluble
materials)

cortex
(parenchyma cells packed
with many dark bodies
– starch grains)

cell
wall

endodermis
(very thick walls
in U-shape)

pericycle
(thick-walled cells
– many layers)

phloem sieve
tube elements

xylem vessel elements
(thick-walled empty
cells)

Smilax herbacea
(carrion flower)
Root, cross section
100 x

pith
(parenchyma cells
with thick walls
extending to
center of root)

drawn unambiguously. Provide a legend so that you know exactly what is being represented; note the magnification for microscope drawings.

Soon after each lab period, go through your notebook, expanding it with additional comments and redrawing illustrations that might confuse you later. Finally, if possible, take full advantage of review labs. See Chapter 2 for additional tips on recording information in the laboratory and the field.

STUDYING FOR SHORT-ANSWER QUESTIONS

Most biology exams include a variety of short-answer questions. For example, there may be multiple-choice, matching, or true/false sections, as well as questions calling for answers of a few words to a few sentences.

Do not simply reread your notes and textbook to study for such tests. Following are ways to use writing productively to make studying a more active experience.

■ Create your own expanded "textbook."

Make one coherent study guide by integrating lecture notes, course readings, and material from laboratory or field sessions. After all, your instructor sees all these components as related to one another; you should, too.

■ Compile a list of key terms.

Look for words appearing in italics or boldface in your text, and check the end of each chapter for a glossary or list of important words. Also, consult your notes for terms that your instructor has emphasized or used repeatedly. Write out the definition to each term in your own words; then check your book or notes for accuracy. Do this repeatedly until you have mastered the important vocabulary.

■ Devise your *own* questions.

Then devise "perfect" answers. File cards are useful here: you can put the question on one side and the answer on the other. Don't simply flip the cards; write out your response to each question again on a separate sheet of paper, and check it against your prepared answer. Repeat the process until you feel confident with the material. This activity is also an excellent strategy for studying for essay exams (see below).

■ Do not become lost in details: focus on central issues.

One of the most difficult tasks for beginning students is deciding at what level to aim their mastery of a subject. In practical terms, this translates into worrying about how picky an instructor is going to be on tests. This problem has no simple solution. However, even when you are aiming for quick recall of a large body of information, focus first on the main points. Biology is not a collection of isolated facts: much more important are the underlying *patterns* and unifying *concepts*. Try grouping related items: three factors that influence a particular process; major similarities between two groups of organisms; six characteristics of a particular family of animals; and so on. Also consider the interrelationships that may exist among important ideas. Some people find it helpful to diagram these connections, starting with a single main idea or topic and surrounding it with related concepts joined by connecting lines. You will remember the particulars if you first understand the generalities.

■ Find a study partner, or create a study group.

Research has shown that many people profit from collaborative learning. Consider working with one or more compatible, motivated persons on a regular basis to revise and review lecture notes and prepare for exams.

ANSWERING ESSAY QUESTIONS

The above advice on preparing for short-answer questions applies as well to essay exams. However, essay questions also pose other challenges. Awkward, wordy, or ambiguous prose will suggest that your thinking is also sloppy and unrefined, and that you do not fully understand the material. Clarity and coherence are as important in essay exams as in other forms of biological writing, yet time constraints usually do not allow much opportunity for rewriting. For this reason, good essays are the product of well-focused studying followed by careful use of time during the examination. These strategies may help you:

■ Read through the entire test before you start.

Check whether you must answer all questions or if you have some choice. Do the *easiest* sections first. Budget your time carefully; give yourself a time limit for each question. Leave a few minutes at the end to review your answers and to expand them if necessary. Check again that you have answered or attempted all required questions.

■ Address the question *asked*.

This may sound obvious; however, many students do not read the question carefully. Some essays call for a straightforward summary of information, whereas others ask you to think more creatively; understand this difference. Focus on key words in the question: *define, list, compare, contrast, evaluate, analyze, describe, discuss*. Look for possible choice *within* questions.

■ Stick to the point.

You will not get a higher grade for adding irrelevant information. Such material obscures the strong parts of your answer and suggests that you don't fully understand the subject. It also wastes valuable time.

■ Observe length restrictions on answers.

Guidelines for the length of answers (one concise paragraph, two pages of the examination booklet, and so on) convey your instructor's advice about the depth of detail needed, and about the amount of time you should reasonably spend on the question. Point values for specific questions also guide you in this respect. If a question is worth 10 points out of 100, avoid spending more than 10% of your time on it — in other words, no more than 5 minutes during a 50-minute exam.

■ Plan before you start to write.

Organize your thoughts by jotting down a brief outline or list of important points. Crucial facts are easy to forget when you are working under pressure. Taking a minute or two to arrange your thoughts will save time later and help you write a more complete answer.

■ Develop a clear thesis.

Essays requiring you to integrate, analyze, or evaluate material can easily lose focus. Do not ramble; organize your answer around a clearly stated central idea. State this in a topic sentence (see p. 132); then build the rest of the essay around your main point. Sometimes the wording of the question gives you raw material for a thesis statement; however, don't waste time unnecessarily repeating the entire question verbatim.

■ Support general statements with specific evidence or examples.

Many essays are inadequate because the writer conveys only a superficial knowledge of the material. Thorough understanding of a subject involves the ability to explain how broad ideas rest on a foundation of supporting details. Here is your chance to display the depth of your preparation.

■ Integrate illustrations with the text of your answer.

If you find it useful to add graphs or drawings to your essay, make sure they are well labeled and clearly related to your discussion. Do not simply tack them on at the end; refer to them specifically in your answer.

Successful essays on examinations are usually the product of much advance preparation. If your studying has been thorough and well focused, you will already have thought of many possible essay questions yourself — perhaps some of the same ones your instructor has devised. In other words, you will have already tried to make connections between topics, identify central issues, and see the course material in new ways. All these are purposes of essay questions.

CHAPTER **10**

Other Forms of Biological Writing

ORAL PRESENTATIONS

Many biologists participate in professional conferences or symposia that offer a forum for exchanging ideas, conveying information about new research methods, and reporting research in progress. Conference proceedings are often available in printed form; however, oral exchange of information is still the primary means of communication.

As a biology student, you may be required to prepare an oral presentation of your own — perhaps a summary of a published paper, a review of the literature on a particular topic, or even a report of your own research for a seminar or a professional conference. Although the final product will not be presented in written form, you will still need to use writing to organize your talk. Actually, there are many similarities between writing a paper and planning an oral presentation. Both activities require you to consider your audience and your purpose and to convey information clearly, accurately, and logically. Both also force you to examine your own understanding of the material and to use writing as a means of clarifying your thoughts.

Following are some practical suggestions for giving successful oral presentations.

■ Write out the entire talk beforehand.

Even if you are an accomplished speaker, putting everything down in writing will make your presentation more organized and coherent and will lessen the chance that you'll forget an important point. Plotting out the talk sentence by sentence also allows you to plan your words carefully and search for the most effective ways of explaining difficult ideas. After you rough out the first draft, revise the manuscript carefully, looking for sections that are poorly worded or likely to confuse the audience, or that jump abruptly from one topic to another.

■ Use an appropriate method of organization.

How you organize the presentation will depend, of course, on the subject and your objectives. Nevertheless, planning will be easiest if you think of the talk as having a distinct beginning, middle, and end. You might start by briefly reviewing general background material, then narrowing down to your specific objectives. Be sure to keep introductory material brief, or else you'll have to rush through the most important points. Save time at the end to summarize, offer conclusions, discuss broader aspects of the topic, and answer questions.

If you are reporting on original research, use the format of a research paper (see Chapter 4) to guide you. Begin with a short literature review, placing your own study in context. Then get specific: Why did you do this study? What did you expect to find? Briefly explain your materials and methods. Next, in the middle section of your presentation, summarize the results. Here, focus on your major findings, adding just enough supporting information (details, statistics, examples) to develop your points. Conclude by looking again at your objectives. Do the data support your original hypothesis? What connections can you make to the findings of other researchers? What questions remain?

■ Never *read* a prepared talk word for word.

Doing so will suggest (perhaps correctly) that you are not comfortable enough with the material to abandon your notes. It will also distance you from your listeners, who will rapidly lose interest in your formal recitation. Instead, use the written version of the talk to make a brief list of key points or concepts; these can be put on a single sheet of paper or on file cards arranged sequentially. As you speak, use the key points to jog your memory and keep you on track. Once you become thoroughly familiar with the talk in its written form, you will probably remember your most effective sentences and phrases word for word and will be able to say these naturally as if they just occurred to you. An effective talk — one that really engages the audience — strikes the proper balance between carefully structured wording, worked out in advance, and a spontaneous, informal delivery.

■ Observe your time limit.

Presentations at professional conferences are subject to strict time constraints: if you run out of time, you may have to stop abruptly even if you're only midway through your talk. Even student presentations must take into account the instructor's plans, the needs of other speakers, and the duration of the class period. Running overtime is not only a sign of poor preparation but also a discourtesy to others.

Practice speaking in front of a friend or rehearse the talk mentally, timing the entire presentation from start to finish. Determine what you should be discussing halfway through the allotted time period, and note this point both on the final draft of the talk and on your speaking notes.

Do not speak too rapidly. If you are nervous you may speak more quickly than usual, especially as you approach the end of the talk. Remember that listeners need time to digest everything they hear. If you confuse them at any point, they may stay confused for the rest of the talk; unlike readers of written text, they cannot easily go back and review difficult sentences. It helps to pause briefly after important points and to repeat difficult material in slightly different wording. If your format permits, you can also invite questions at potentially confusing places in the presentation.

■ Do not swamp listeners with details; avoid jargon.

Develop only a few main points. Even highly attentive listeners can take in only so much information at one time; they'll lose track of your argument if you bombard them with too many details. Use clear, straightforward language, and avoid jargon. (For more on writing clear, accurate sentences, see Chapter 7, pp. 135–43.) Tailor your talk to your *audience*. Explain any terms likely to be unfamiliar to listeners, but keep the number of such terms to a minimum.

■ Establish eye contact with the audience.

Doing so will make you feel more relaxed when speaking and will make your audience more receptive.

■ Use visual aids, if possible.

In the classroom, the blackboard is a traditional and still invaluable tool even for short presentations. Take advantage of it as you talk to write down unfamiliar terms or important statistics and to draw simple tables and graphs. Your writing should be large enough to be seen clearly from the back of the room. Graphs should be simple to read and clearly labeled. Do not clutter the board with disorganized scrawls, and do not inadvertently erase information you must refer to frequently.

Material that is relatively time-consuming to draw should be put on the blackboard *before* your talk; otherwise you'll waste time writing while the audience waits impatiently or struggles to copy what you have written. It may make more sense to distribute such material as a supplementary handout that the audience can look at as you talk. Handouts are also useful for listing key terms and definitions, important points to be covered, or useful references. However, do not make handouts too detailed or extensive, or people will spend their time reading instead of listening to you.

If you are presenting an original research project, you may wish to show slides illustrating various aspects of the study, such as the field site, organisms studied, experimental apparatus, and quantitative data in the form of tables or graphs. Slides can also be used to display important terms or a short list of main points. However, to be effective, slides must be of good quality. Do not make the common mistake, seen even at professional meetings, of showing poorly exposed, blurry, or otherwise unsuitable slides accompanied by repeated apologies ("Sorry these slides are so dark, but . . ."). Such slides are not worth using at all. If you are not a competent photographer, check the facilities of the audiovisual department at your institution, or think of an alternate method for illustrating your talk.

If you do have sharp, uncluttered, well-exposed slides, spend time planning how you will integrate them with the text of your talk. As in a written paper, each illustration should clarify or support a particular statement. Write out beforehand what you plan to say about each slide, and omit any slides that serve no clear purpose. Write down in your lecture notes when each slide should be shown. Budget your time carefully to take into account how much time you will devote to each slide. When possible, try to show slides in one or more groups; otherwise, you will distract the audience by repeatedly turning the lights on and off. Set up the projector and screen in advance, locate a pointer, and check that all slides are the right side up and in the proper order.

Slides depicting numerical data should be simple enough for the audience to grasp quickly. Do not pack them with more information than people can absorb in a minute or two. Axes of graphs should be clearly labeled, and all words and numbers should be large enough for viewers to read on their own. Remember that slides showing data should generally stress concepts, not particulars. Use them carefully and sparingly to make your points.

Overhead transparencies of figures and tables are also effective lecture illustrations. Consider using blank transparent sheets to jot down terms or make quick drawings as you talk. You can also prepare some transparencies in advance. An advantage of overhead projection is that it allows the speaker to leave the room lights on and thus to interact more directly with the audience.

Finally, you may wish to take advantage of the versatility of Powerpoint (Microsoft Corp.), which allows you to mix text with graphs, tables, photographs, or videos, including animation. Many of the same guidelines for slide presentations apply also to those using Powerpoint. Remember that your computer must be compatible with the projection system you will be using. It is wise to check the setup well in advance.

■ Be prepared for questions.

You cannot predict everything you will be asked, but you probably can anticipate some of the questions. Write them out beforehand and prepare brief, concise answers. If you are asked a question for which you are unprepared, do not try to bluff your way through a reply. It is far better to say that you don't know the answer. If you have given a thoughtful, well-organized talk, listeners will already be convinced that you know your subject. They will not expect you to know everything.

POSTER PRESENTATIONS

A poster presentation conveys an author's original, unpublished findings *visually,* through a selective assemblage of illustrations (such as graphs, tables, and photographs) that are carefully integrated with a small amount of text. All these materials are fastened to a poster or display board. Posters have become popular forms of communication at scientific conferences, and considerable space may be set aside for them. Often there are designated time periods for the authors of posters to be on hand to answer questions. Some people prefer giving posters to delivering talks because they feel more comfortable with the format; in other cases, a particular study may lend itself more readily to a poster presentation.

From the audience's perspective, browsing through a poster session at a busy conference offers a pleasant interlude during a tightly scheduled sequence of talks. In addition, because oral presentations are generally brief, with an even shorter question-and-answer period, poster sessions allow for more extended, informal discussions between a researcher and those interested in his or her work.

If you are considering a poster presentation of your own study, the following advice may help you get started.

■ Follow conference guidelines closely.

Study the conference circular or Web site for details about poster size, contents, and other requirements. Stands and other basic materials are often supplied to presenters, so it may be easiest to bring with you the various components of the display and assemble them on site. Remember, too, that posters, like orally delivered papers, must be formally accepted for presentation by the organizing or editorial committee of the conference. Prospective participants are usually asked to submit a title and an abstract. It makes little sense to spend large amounts of time preparing a poster until, first, it has been accepted, and second, you have a clear sense of what is actually required.

■ Use a simple and logical organization.

In many ways, the format of a poster can be similar to that of a written or orally delivered paper. Start with a specific statement of the scientific problem, followed by brief mention of materials and methods, presentation of main results, and a short, concise discussion or conclusions section. If you cite other literature, do so sparingly. Your design should invite the reader to start at one place and proceed logically according to a particular sequence. For example, materials can be arranged on the display board so that they read from left to right, or from top to bottom. Do not forget the importance of a specific (and compelling) title.

■ Focus on the main points. Remember both the advantages and the constraints of a visual format.

Most readers of posters don't have the patience or the interest to stand around reading dense chunks of text. Keep written portions to a *minimum*. A poster should be highly visual with an attractive, unclut-tered appearance. Focus on the most important ideas, and present them as simply as possible. Supporting points and details will have to be saved for informal discussion and, later, for publication. Use graphs and tables to highlight only the major points, trends, or relationships — not to pack in lots of data. Headings, labels, numbers, and text should be easily readable from at least several feet away. Consider using lists of numbered or bulleted points, rather than paragraphs of text, to summarize results or conclusions. Don't forget the potentials of photographs or other types of illustrations. Finally, unless you have access to the proper equipment and are skilled in using it, consider getting professional help in preparing your visual materials.

■ Prepare in advance for questions and informal discussion.

As with an orally delivered paper, try to anticipate some of the questions you may be asked, and prepare for them by writing out notes or composing brief replies. You may also wish to prepare handouts containing more details on your study, as well as your name and academic address. Some authors of posters also bring reprints of papers they have written on similar topics. If your poster is well prepared, it will attract others working in your subject area and may even engage people with peripheral interests. You may learn a great deal from thoughtful readers of your poster.

RESEARCH PROPOSALS

Research proposals have many traits in common with research papers (see Chapter 4): both introduce a scientific question or hypothesis, both put a specific study in the broader context of existing research, and both are organized in ways that reflect the logic of the scientific method. Of course, a research proposal is missing a Results section, along with the author's analysis and interpretation of those results. Instead, it seeks to propose and justify the author's research *plans* — perhaps for a senior project, a summer internship, a graduate thesis, or, in the case of professional biologists, new or continuing work in a specialized field.

There is no one standard format for a research proposal. Authors seeking funding from the National Science Foundation or other agencies must adhere to the precise guidelines laid down by those organizations; student authors should follow their instructor's or university's requirements. Nevertheless, all research proposals involve some of the same challenges and constraints. If you are presented with the opportunity to do research, the following guidelines should help you in preparing your own proposal.

■ Create a specific and informative title.

A focused title suggests a well-thought-out project. Titles such as "A Study of Stream Ecology" or "Nutrient Uptake in Seaweeds" convey little sense of direction or purpose; compare them with more specific versions: "The Effect of Open and Close Canopy on the Diversity of Macroinvertebrates in Morrisville Creek"; "Ammonia Uptake by *Ulva curvata* Growing on Three Different Substrates." You may need to modify your title as you proceed with the research, but a good working title is essential to a well-received proposal.

■ Start with an overview of the general scientific issue.

Before proposing your own study, you need to introduce the broader topic to which your work relates. Most research proposals begin with an Introduction (or Background to the Study) section, which presents a theoretical context for the proposed investigation by summarizing and integrating the work of others in the field. Doing a good job on this part involves a thorough literature search and careful, selective use of primary sources, which are cited in your text and then listed later in a Literature Cited section. Your introductory discussion should also reflect a sensitivity to the scientific background possessed by your readers. Typically, you'll be submitting your proposal to other scientists; however, if it will be judged by nonscientific readers (for example, a multidisciplinary committee), then keep specialized vocabulary to a minimum, and use a less technical style.

■ State the aims of your study and provide a clear rationale.

After providing background material and highlighting major findings, you can then turn to important gaps in our knowledge. What conflicts exist in the literature? What questions remain? Relate such issues to your *own* proposed study and its specific aims. What questions will you investigate? How will your work enlarge, clarify, or complement existing knowledge of the subject? Discussion of your own objectives can come at the end of the Introduction or Background section, or in a subsequent section.

■ Summarize your methods.

Next explain specifically how the research will be conducted, what kinds of data you will collect, and how those data will be analyzed. The level of detail, and thus the amount of space devoted to the Methods section, will depend on the type of proposal you are submitting and the audience for whom it is intended. Remember that even the most promising research question will fall flat if you can't devise a practical and scientifically valid way to investigate it. Think through your methods completely, and become thoroughly familiar with the procedures used in studies similar to yours. If you want to convince your readers to commit their time and/or money to your project, then it must sound feasible.

■ Provide a budget, if necessary.

Here, too, follow whatever specific guidelines apply to your situation. If an estimate of projected expenses is requested, prepare this with painstaking care. You will need to identify potential suppliers, compare prices, and compile a detailed list. Don't be vague: "collecting supplies — $100"; instead, give a breakdown of exactly what you'll need and how much each item will cost: "1 dip net, $38.50; 2 replacement bags @ $8.98 each; 1 D-frame aquatic net, $39.95. . . ." This task may take much time and effort; however, it will ensure that you consider your methodology very carefully. Moreover, your proposal will more likely be accepted if you show that you can do what you propose within the budget allotted to you.

■ Follow application instructions carefully and submit your proposal on time.

This advice may seem obvious, yet many applicants run the risk of having their proposals rejected out of hand because they fail to include all required materials. Pay meticulous attention to such requirements as page length, format, margins, spacing, and the number of copies to be submitted. Check to see if your résumé and/or letters of recommendation are also required. You may also need to prepare a brief cover letter to accompany your proposal. Finally, be prepared to meet the application deadline; failure to do so suggests immediately that you are incapable of adequate planning and a professional approach.

Following is a sample research proposal for an undergraduate summer research fellowship.

Compose a specific and informative title. Include your name and other relevant information on a title page.

Field Measurements of
Photosynthesis and Transpiration
Rates in Dwarf Snapdragon
(*Chaenorrhinum minus* Lange):
An Investigation of Water Stress
Adaptations

Tara Gupta

Application for Summer Research
Fellowship,
Colgate University
March 12, 200-

Water Stress Adaptations 2

INTRODUCTION

Dwarf snapdragon (*Chaenorrhinum minus*) is a weedy pioneer plant found growing in central New York during spring and summer. Interestingly, the distribution of this species has been limited almost exclusively to the cinder ballast of railroad tracks,[1] a harsh environment characterized by intense sunlight and poor soil water retention. Given such environmental conditions, one would expect *C. minus* to exhibit anatomical features similar to those of xeromorphic plants (species adapted to arid habitats).

However, this is not the case. T. Gupta and R. Arnold (unpublished) have found that the leaves and stems of *C. minus* are not covered by a thick, waxy cuticle, but rather with a thin cuticle that is less effective in inhibiting water loss through diffusion. The root system is not long and thick, capable of reaching deeper, moister soils; instead, it is thin and diffuse, permeating only the topmost (and driest) soil horizon. Moreover, in contrast to many xeromorphic plants, the stomata (pores regulating gas exchange) are not found in sunken crypts, or cavities in the epidermis that retard water loss from transpiration.

Despite a lack of these morphological adaptations to water stress, *C. minus* continues to grow and reproduce when morning dew has been its only source of

Introduce the scientific issue and give background information. Cite relevant studies by others; number system is illustrated here. Adjust vocabulary to take into account the biological background of your readers.

Water Stress Adaptations 3

water for up to five weeks (R Arnold, personal communication). Such growth involves fixation of carbon by photosynthesis, and requires that the stomata be open to admit sufficient carbon dioxide. Given the dry, sunny environment, the time required for adequate carbon fixation must also mean a significant loss of water through transpiration as open stomata exchange carbon dioxide with water. How does *C. minus* balance the need for carbon with the need to conserve water?

AIMS OF THE PROPOSED STUDY

State aims and scope of proposed study.

The above observations have led me to an exploration of the extent to which *C. minus* is able to photosynthesize under conditions of low water availability. It is my hypothesis that *C. minus* adapts to these conditions by photosynthesizing in the early morning and late afternoon, when leaf and air temperatures are lower and transpirational water loss is reduced. During the middle of the day, its photosynthetic rate may be very low, perhaps even zero on hot, sunny afternoons. Similar diurnal changes in photosynthetic rate in response to midday water deficits have been described in crop plants.[2,3] There appear to be no comparable studies on noncrop species in their natural habitats.

Thus, the research proposed here should help explain the apparent paradox of an organism that thrives in water

Water Stress Adaptations 4

stressed conditions despite a lack of morphological adaptations. This summer's work will also serve as a basis for controlled experiments in a plant growth chamber on the individual effects of temperature, light intensity, soil water availability, and other environmental factors on photosynthesis and transpiration rates. These experiments are planned for the coming fall semester.

METHODS

Simultaneous measurements of photosynthesis and transpiration rates will indicate the balance *C. minus* has achieved in acquiring the energy it needs while retaining the water available to it. These measurements will be taken daily at field sites in the Hamilton, NY, area, using an LI-6220 portable photosynthesis system (LICOR, Inc., Lincoln, NE). Basic methodology and use of correction factors will be similar to that described in related studies.[4-6] Data will be collected at regular intervals throughout the daylight hours, and will be related to measurements of ambient air temperature, leaf temperature, relative humidity, light intensity, wind velocity, and cloud cover.

Briefly describe your methodology.

BUDGET

1 kg soda lime, 4-8 mesh $70
   (for absorption of CO_2 in photosynthesis analyzer)

If a budget is required, be as specific as possible.

Water Stress Adaptations 5

1 kg anhydrous magnesium perchlorate $130
   (used as desiccant for photosynthesis analyzer)

SigmaScan software (Jandel Scienti- $195
fic Software, Inc.)
   (for measurement of leaf areas for
   which photosynthesis and transpiration rates are to be determined)

Estimated 500 miles travel to field $140
   sites in own car @ $0.28/mile

CO_2 cylinder, 80 days rental $20
   @ $0.25/day
   (for calibration of photosynthesis
   analyzer)

                              TOTAL REQUEST $555

REFERENCES

1. Wildrlechner MP. Historical and pheno-
   logical observations of the spread of
   Chaenorrhinum minus across North Amer-
   ica. Can J Bot 1983;61:179-87.

2. Boyer JS. Plant productivity and environment. Science 1982;218:443-8.

3. Manhas JG, Sukumaran NP. Diurnal
   changes in net photosynthetic rate in
   potato in two environments. Potato Res
   1988;31:375-8.

4. Doley DG, Unwin GL, Yates DJ. Spatial
   and temporal distribution of photosynthesis and transpiration by single
   leaves in a rainforest tree, *Argyro-
   dendron peralatum*. Aust J Plant Physiol 1988;15:317-26.

5. Kallarackal J, Milburn JA, Baker DA.
   Water relations of the banana. III.

Include all pub-
lished works
cited. Numbers
correspond to
order in which
sources were
first mentioned
in the text.
CBE citation-
sequence sys-
tem is shown
here. Author's
last name is fol-
lowed by ini-
tials, then paper
title, journal,
publication
date, volume,
and page num-
bers.

Water Stress Adaptations 6

Effects of controlled water stress on water potential, transpiration, photosynthesis and leaf growth. Aust J Plant Physiol 1990;17:79-90.

6. Idso SB, Allen SG, Kimball BA, Choudhury BJ. Problems with porometry: measuring net photosynthesis by leaf chamber techniques. Agron 1989;81: 475-9.

LETTERS OF APPLICATION

Letters of application, along with cover letters accompanying résumés, are never mere formalities. Prospective employers or graduate school admissions committees do read such letters, often before looking at any other material about you. If you want to create a good first impression, then spend as much time as possible on this stage of the application process. Draft *each* cover letter carefully, revise it thoughtfully, and proofread it meticulously. Use the letter to grab the reader's attention, identify yourself as a candidate, highlight your best qualities, and demonstrate a sincere interest in the position for which you are applying. A compelling cover letter will distinguish you from the pool of other applicants; it also may offset, in the reader's mind, any deficiencies in your qualifications. The following guidelines should help you.

■ Use a standard business letter format.

One variation of this format is illustrated in the sample letter on page 186. Using a computer, type your return address (omitting your name), followed by the date, in the upper right-hand corner; then space down three or four lines, and at the left margin type the name and address of the person to whom you are sending the letter. Double-space; then type the salutation (Dear Mr. _____, Ms. _____, Dr. _____, or other appropriate greeting) followed by a colon. Double-space again and begin the body of your letter, single-spaced, with double-spacing between paragraphs (optional). Indent each paragraph five spaces. After the last paragraph, double-space and type a standard closing (Sincerely, Yours truly, or the like), followed by a comma; then skip four lines and type your name. The closing and your name are placed at the right-hand side of the letter, in line with the return address. In the space between the closing and your typed name, add your written signature, using your full name.

A variation is to type all components of the letter (return address, date, closing, your name, and so on) flush with the left margin. If you choose to prepare your cover letter this way, you do not indent paragraphs (but be sure to double-space between them).

■ Print the letter on appropriate stationery.

Do not write out the letter in longhand. Use good-quality, unlined paper — 8½ by 11 inches is a standard size; never use informal personal stationery. Print on only one side of the page. Remember that your cover letter should convey a sense of seriousness and professionalism.

■ Be straightforward and concise.

Restrict your letter to a single page, if possible. Your reader may be screening a large number of applicants and may have neither the time nor the patience to deal with a long treatise about you and your many qualifications. Remember the purposes of a cover letter: to identify yourself as an interested candidate and to highlight your strengths. There is no need to paraphrase your entire résumé.

■ Use specific language and details.

Experienced readers can quickly spot the form-letter approach used by some applicants to save time and effort. This tactic involves composing a standard, general cover letter and sending it out (often along with a standard, general résumé) to a large number of addresses. Consider the following example:

```
Dear Department Chair:

        I am a senior biology major at Shoreham
College. I find your university very interest-
ing and would like to apply to the Ph.D. pro-
gram. Please also consider me for a teaching
assistantship. Enclosed is my résumé.
        Thank you.

                           Yours truly,
```

Form-written cover letters often do you more harm than good. If you want to stand out from other applicants and convince your reader that you are sincerely interested in this *particular* job or opening, then you must individualize each letter. What is it about this specific university, program, business, or agency that draws you to it? What special aspects of your background or training suit you for this opening? What makes you a unique applicant for the position? Do not speak in vague generalities: shape your letter to fit the context, and support your points with specific details. Strive for a positive, confident tone without sounding boastful. Of course, you must write honestly and sincerely; do not say anything that you may have to back down from later.

Wherever possible, send your letter to the person who is *directly* involved in the hiring or acceptance process; otherwise, your letter may get sidetracked. If you do not have this person's name, call the business or university to obtain it. Be sure to check the spelling of the person's name to whom you are sending your letter.

The following sample is a letter of application for a research assistant position. Note how the writer has tailored her letter to a particular reader and to the specifics of the job situation.

Box 7337
Hamilton University
Old Westbury, New York 11568
April 17, 2000

Professor William Rider
Biology Department
Southwestern University
Tucson, Arizona 85721

Dear Professor Rider:

I am a junior biology major at Hamilton University, with a strong interest in plant ecology. I am particularly interested in the reproductive strategies of desert plants, and hope eventually to do research of my own in this area. While researching a paper for an ecology seminar this semester, I became interested in your work on flowering and seed dispersal in *Opuntia* spp. I understand that you regularly employ several undergraduate students as research assistants every summer. I am writing in the hope that I might be considered for one of these positions.

Although I grew up in the Northeast, I have relatives in Tucson whom I visit frequently. During one visit, when I was sixteen, I attended a weeklong desert camping trip for high school students, sponsored by Southwestern University's Extension Program. It was this experience, combined with a passion for hiking and an amateur interest in nature photography, that first sparked my interest in the desert. Over the next few years, I developed a good field knowledge of desert flora through camping trips in Saguaro National Monument and Organ Pipe Cactus National Monument. Although I have not taken any courses similar to your Biology 341 (Desert Ecology), I have taken both Introductory Ecology and the Advanced Ecology Seminar at Hamilton University, as well as several botany courses. In addition, last summer I worked for Dr. Janet Rith-Najarian as a research assistant, gaining experience with plant sampling techniques in local deciduous forests. I have a good background in the taxonomy of vascular plants, as well as considerable experience with data analysis using computer software packages.

If accepted for a position, I could arrange to be in Tucson by May 25, 2000, and would be available for full-time employment through August 20. I will be at Hamilton University until May 18, after which I can be reached at home (8 Hicks Lane, St. James, New York 11780). My home phone number is 555-825-7713.

Thank you for considering me as an applicant. I look forward to hearing from you.

Sincerely,

Marin Johnson

Marin Johnson

RÉSUMÉS

Composing an effective résumé may be one of the most important writing projects you'll face during your academic or professional career. If carefully crafted, your résumé may be instrumental in getting you that first interview and a job, or acceptance in a graduate program. If poorly prepared, it may generate little response from its readers, or even elicit a strongly negative reaction, damaging your chances to be considered further.

Your basic task in writing a résumé is to promote yourself as a candidate by highlighting your most important achievements and qualifications. A résumé usually begins with your name, address, and phone number, followed by such information as your career objectives, educational background, employment history, research and teaching experience, awards received, and other relevant interests, skills, or activities. You are not required to give personal information such as sex, race, religion, marital status, or age. There is no one right way to put all this material together; however, the following suggestions will apply to the preparation of most résumés:

■ Use a professional-looking and inviting format.

Your résumé should be typed on a computer and printed on good-quality paper to match your cover letter. Organize information using headings (and, if necessary, subheadings); the sample résumé on page 190 illustrates a common format. Notice the use of single- and double-spacing to tie together related material and to separate successive sections. Use capitalization, boldface type, underlining, indentation, or other devices to organize your text, but be sure to be consistent. A chronological approach is often used to list related information, such as jobs held or degrees earned, starting with the most recent items and working backward. However, do not feel bound to this convention. If, for example, work you did five years ago is most relevant to the position you want, then list this employment first, followed by other work of less importance.

■ Modify your résumé to fit each application.

Most of the material in your résumé will be relatively fixed — for example, your educational history, much of your work experience, and so on. However, think carefully about each position for which you are applying. Some skills and experiences may be more relevant in some situations than in others. Look again at the sample résumé. The fact that the applicant is fluent in Japanese probably has little bearing on her chances for a research assistant job in Arizona; however, if she were applying to a student exchange program in Tokyo, she would do well to expand a description of this skill. In addition to the camp counselor work already listed, she might add that she studied Japanese for three years in high school and two years at college, and also spent three weeks in Japan one summer. Similarly, the cover letter accompanying her résumé (see p. 186) should highlight these and other pertinent parts of her background.

Thus, to tailor your résumé to fit each position, you need to think creatively about yourself. It's easy to overlook (or to dismiss as unimportant) such experiences as unpaid work at a school or hospital, membership in special-interest clubs, athletic or musical abilities, evening or weekend classes, relevant hobbies, and so on that might be of interest. Put yourself in the position of your prospective employer, school, or program, and try to imagine the "package" of qualities you might look for in a candidate. Your résumé should present your particular version as convincingly as possible.

■ Be concise and selective.

Remember that you'll probably be competing with many other applicants; consequently, your readers may take only a quick glance at your credentials. In any event, most people will be put off by a lengthy résumé unless every item in it is pertinent and noteworthy. Never pad your résumé with unnecessary details or other material just to fill up space; experienced readers will quickly suspect that you are trying to look better on paper than you really are. A good résumé is straightforward, succinct, and short — typically a single page. On the other hand, a curriculum vitae, which contains more detailed information about your coursework, field experiences, lab skills, and other academic qualifications, may be two pages or longer. You will probably want to prepare a curriculum vitae rather than a résumé if you are applying to graduate school or seeking fellowships or other academic opportunities.

■ Update your résumé frequently.

Add new accomplishments, skills, and experiences as you acquire them; at the same time, modify or delete outdated material.

■ Edit and proofread meticulously.

Even though you will be working mostly with lists and sentence fragments, choose your words carefully. Use parallel grammatical forms for lists of similar items. Remember that even a single misspelled word or typographical error will detract from the professional appearance you want to convey. Give your résumé to other people to proofread (and critique). Try reading it aloud to spot errors.

■ **Supply references.**

This may be done in several ways. Your résumé can include the names, addresses, and phone numbers of several people who may be contacted for letters of recommendation. Another option is to give this information in your cover letter. A third method is simply to say "references available on request." You will then be contacted for more information if your prospective employer or program is interested in your application. Be sure to check the application instructions or employment advertisement to see if there are special guidelines about how or when references should be supplied.

Also, of course, you'll need to obtain the permission of each person whom you wish to list as a reference. Remember that such people should be familiar with at least some of your abilities and strengths; good choices include former employers, and instructors in whose courses you distinguished yourself. Give these persons plenty of advance notice — at least a few weeks — and make things easier for them by supplying a copy of your résumé, information about the position, and an addressed, stamped envelope to use in mailing their recommendation. Be sure to inform them of the application deadline. It's also a courteous gesture to follow up with a brief note of thanks, whether or not your application was successful.

The following sample student résumé (p. 190) should give you some idea of what a well-crafted and well-prepared résumé should look like. See also the sample curriculum vitae on pages 191–92.

SAMPLE RÉSUMÉ

MARIN JOHNSON
8 Hicks Lane
St. James, NY 11780
555-825-7713

GOALS
: Graduate work, followed by a teaching and research position in plant ecology.
Special interest in desert flora.

EDUCATION
: Hamilton University, Old Westbury, NY. Bachelor of science expected May 1996.
Major in biology.

RESEARCH EXPERIENCE
: Research assistant in plant ecology for Dr. Janet Rith-Najarian, Biology Department, Hamilton University, summer 1994.
Assisted with plant collections, species identification, and data analysis.

TEACHING EXPERIENCE
: Teaching Assistant, Biology 101, Hamilton University, spring 1995.
Took major responsibility for setting up labs; also conducted review sessions and assisted with grading.

Instructor, Outdoor Photography Workshop, Madison County Nature Preserve, Poolville, NY, June 12–13, 1994.
Devised and taught own 10-hour course for 23 adults.

Writing tutor (sciences), Interdisciplinary Rhetoric Department, Hamilton University, spring 1993.
Instructed first-year students on the preparation of lab reports for introductory biology and chemistry courses.

ADDITIONAL WORK EXPERIENCE
: Camp counselor, Experiment in International Living (Japan division), Otselic Lake, MN, summer 1993.

Clerk, Stony Brook Natural Foods, Laurel, NY, summer 1992.

Volunteer guide, Madison County Nature Preserve, Hubbardsville, NY, summer 1991.

HONORS
: Chandler Writing Prize (for best undergraduate essay), Hamilton University, May 1994.

Photograph of the Year Award, *Outdoor Life* magazine, 1993.

SPECIAL SKILLS AND INTERESTS
: Competent nature photographer.

Experienced in wilderness camping, especially in Arizona deserts.

Fluent in Japanese.

Member of swim team, 1993–1994.

REFERENCES
: Available on request.

ANDREW D. ROZBRUCH

3 Haydn Avenue	Colgate University
Syracuse, New York 13210	Hamilton, New York 13346
555-422-1766	555-824-1000
	adr@center.colgate.edu

OBJECTIVE

Entry-level position in a clinical or research lab

EDUCATION

Colgate University, Hamilton, New York   Class of 2000
    Bachelor of Arts with major in Biology   GPA: 3.6

University of New South Wales, Sydney, Australia, fall semester, 1999

HONORS

Dean's Award for Academic Excellence: 1997–2000
Tri-Beta National Biological Honors Society: Inducted October 1998

COURSEWORK

General Biology, Vertebrate Environmental Physiology, Ecology, Conservation Biology, Plant Ecosystem Processes, Genetics, Cellular Biology, Senior Research in Biology, Writing in the Natural Sciences, Chemistry I and II, Calculus, Statistics

SCIENTIFIC KNOWLEDGE

Research Experience

    Senior Research Project, Colgate University. Performed a semester-long study for course credit on the longevity of rhesus monkeys in relation to caloric restriction. Spring 2000.

    Research Assistant, Clinical Research Department, Long Island Jewish Medical Center, Lake Success, NY. Participated in laboratory research on Lyme disease. Responsibilities included running Western Blot, ELIZA, and PCR tests. Summer 1997.

Healthcare Experience

    Health Sciences Intern, Hilton Family Practice, Hilton, NY. Shadowed family physician and gained exposure to the fields of pediatrics, gerontology, OB-GYN, and general medicine. Spring 2000.

    Health Sciences Intern, Johnson Orthopedic Surgery Group, Astor, NY. Observed orthopedic surgery and clinical patient care. Gained an exposure to the field of orthopedic surgery and an appreciation for the doctor-patient relationship. Fall 1998.

    Internship in Dental Clinic, Long Island Jewish Medical Center, Lake Success, NY. Participated in rotation program shadowing and assisting general dentists, oral surgeons, periodontists, and orthodontists. Summer 1998.

Internship in Clinical Research Department, Long Island Jewish Medical Center, Lake Success, NY. Screened patient records in the emergency room for in-house studies. Assisted in obtaining case histories and enrolling patients in studies. Participated in hospital rounds and shadowed research coordinators. Summer 1997.

Teaching Experience

Teaching Assistant and Tutor for Biology 201–Ecology.
Demonstrated laboratory techniques and held office hours to answer questions from students. Fall 1999.

Teaching Assistant for Biology 210–Vertebrate Environmental Physiology.
Assisted with statistical analysis of class data, demonstrated techniques and use of equipment, and held office hours. Fall 1998.

Laboratory Experience

Trained in the usage of various laboratory equipment.
Able to operate a microscope, pH meter, centrifuge, autoclave, spectrophotometer, chromatography machine (thin layer, column, GC), and electrophoresis machine (agarose).

WORK EXPERIENCE

Resident Advisor, Colgate University Residential Life Staff.
Performed dorm rounds, enforced university alcohol and drug regulations, and held educational and social programs for residents. Fall 1998.

Environmental Consultant, Citizens Campaign for the Environment, Utica, NY.
Assisted a nonprofit environmental organization in educating the public on environmental issues. Raised money to support pending environmental legislation efforts. Summer 1999.

Assistant to Director of Planning and Performance, Cablevision, Farmingdale, NY.
Assisted director in planning and preparation of events at Madison Square Garden and Radio City Music Hall. Dec. 1998–Feb. 1999.

ACTIVITIES

Volunteer, Colgate University Safe Rides Program.
Drove courtesy van for Colgate students from bars and fraternity parties to campus. 1999–2000.

Mentor, Pre-Medical Mentoring Program.
Advised prospective pre-medical students about relevant classes and extracurricular activities. 1999–2000.

Senior Class Agent, Senior Class Gift Committee.
Participated in organizing fundraisers and programs to raise money for Senior Class gift to the university. 1999–2000.

Volunteer, Sidekick Program–Big Brother.
Served as a role model to an elementary school student two afternoons a week. Fall 1997–spring 2000.

Brother, Kappa Delta Rho Fraternity, fall 1997–present. Philanthropy Chair, spring 1998.

Rower, Colgate University Crew Team. Fall 1996–fall 1997.

Additional Readings

General Writing Handbooks

Every writer should have at least one general writing guide. The following handbooks cover the essentials of grammar, punctuation, and word choice as well as the crafting of sentences, paragraphs, and whole essays. All are organized as self-help manuals presenting easy-to-follow guidelines applicable to a wide variety of writing situations.

Baker S. 1998. The practical stylist. 8th ed. New York: Longman. 268 p.

Hacker D. 1998. The Bedford handbook. 5th ed. Boston: Bedford/ St. Martin's. 788 p.

Hacker D. 1999. A writer's reference. 4th ed. Boston: Bedford/ St. Martin's. 400 p.

> A concise and easy-to-use handbook that offers thorough coverage of a range of topics, including composition and revision, grammar and mechanics, and research and documentation.

Hairston MC. 1998. Successful writing. 4th ed. New York: Norton. 246 p.

Strunk W, White EB. 2000. The elements of style. 4th ed. Needham Heights (MA): Allyn & Bacon. 105 p.

Weidenborner S, Caruso D. 2001. Writing research papers: a guide to the process. 6th ed. Boston: Bedford/St. Martin's.

Guides to Scientific Writing

The references that follow are particularly useful for more experienced scientific writers and professional biologists, and for undergraduate and graduate students preparing a paper for a technical journal.

The Chicago manual of style. 14th ed. 1993. Chicago: Univ of Chicago Pr. 923 p.

> A widely used reference for writers in many academic fields. Covers a wide range of topics, including documentation methods, tables and figures, use of quotations, and many aspects of writing mechanics and style. Useful as a supplement to guides specifically devoted to scientific writing.

Council of Biology Editors Style Manual Committee. 1994. Scientific style and format: the CBE manual for authors, editors, and publishers. 6th ed. New York: Cambridge Univ Pr. 704 p.

This standard and indispensible reference for biologists contains detailed instructions for preparing scientific manuscripts for publication. The latest edition covers the physical sciences and mathematics, as well as the life sciences. Topics include punctuation, grammar, prose style, manuscript format, tables and figures, chemical notation, cells and genes, symbols and abbreviations, documentation methods, scientific nomenclature, and many others. The manual can be found in the reference section of your library. To order your own copy, contact the Council of Science Editors, Inc., c/o Drohan Management Group, 11250 Roger Bacon Drive, Suite 8, Reston, VA, 20190-5202 (phone: 703-437-4377, fax: 703-435-4390) or visit the Web site at <http://www.councilscienceeditors.org>.

Day RS. 1998. How to write and publish a scientific paper. 5th ed. Phoenix: Oryx. 296 p.

Written with style and wit, this short, readable book gives much practical advice about preparing a paper for publication. It also includes chapters on preparing book reviews, conference reports, theses, and oral presentations. The latest edition contains new chapters on electronic publishing formats, the Internet, and electronic journals.

International Committee of Medical Journal Editors. 1993. Uniform requirements for manuscripts submitted to biomedical journals. JAMA 269:2282–6.

Iverson CI, Flanagin A, Fontanarosa PB. 1998. American Medical Association manual of style. 9th ed. Baltimore: Williams & Wilkins. 672 p.

Woodford FP, editor. 1999. How to teach scientific communication. Bethesda: Council of Biology Editors, Inc., Committee on Graduate Training in Scientific Writing. 199 p.

Although written primarily for instructors of scientific writing, this manual is also useful for both beginning and advanced students. The book contains thorough coverage of the research paper as well as sections on tables and figures, research proposals, theses, and oral presentations. It can be ordered from the publishers of *Scientific Style and Format: The CBE Manual for Authors, Editors, and Publishers* (see above).

Guides to Scientific Literature

Davis EB. 1996. Guide to information sources in the botanical sciences. Littleton (CO): Libraries Unlimited. 275 p.

Hurt CD. 1998. Information sources in science and technology. 3rd ed. Englewood (CO): Libraries Unlimited. 346 p.

An excellent (and very current) source of information and reference guides in all scientific fields.

Roper FW, Boorkman JA. 1994. Introduction to reference sources in the health sciences. 3rd ed. Lanham (MD): Scarecrow Pr. 312 p.

A guide to research in the health sciences divided into three main sections: the reference collection, bibliographic sources, and information sources.

Statistics
The following texts are good introductions for beginners:

Bailey NTJ. 1995. Statistical methods in biology. 3rd ed. New York: Cambridge Univ Pr. 269 p.

Glanz SA. 1997. Primer of biostatistics. 4th ed. New York: McGraw-Hill. 473 p.

The texts below, written for advanced readers, give a comprehensive, thorough treatment of many statistical topics.

Sokal RR, Rohlf RJ. 1994. Biometry. 3rd ed. New York: Freeman. 880 p.

Zar JH. 1999. Biostatistical analysis. 4th ed. Englewood Cliffs (NJ): Prentice-Hall. 929 p.

Figures and Tables
Cleveland WS. 1994. The elements of graphing data. Boca Raton (FL): CRC Pr. 323 p.

For more advanced readers. Covers a wide variety of graphs, with numerous "good" and "bad" illustrations from scientific and technical literature.

Scientific Illustration Committee, Council of Biology Editors. 1988. Illustrating science: standards for publication. Bethesda: Council of Biology Editors. 296 p.

This reference, written for professional biologists and advanced biology students, sets forth guidelines for the preparation of graphs, tables, and other types of visual materials in scientific articles and books.

Tuste ER. 1992. The visual display of quantitative information. Cheshire (CT): Graphics Pr. 197 p.

Guides to Documenting Electronic Sources
The Chicago Manual of Style and the CBE style manual (see pp. 193–94) offer some advice on documenting electronic sources. For more detailed guidance, consult the following manuals or their Web sites.

Crane N, Li X. 1996. Electronic styles: a handbook for citing electronic information. 2nd ed. Medford (NJ): Information Today, Inc. 214 p.

Hacker D. 1999. Research and documentation in the electronic age. Boston: Bedford/St. Martin's. 211 p.

This guide provides an introduction to the Internet with detailed advice for evaluating electronic sources and guidelines for documenting print and online sources. The online version, available at <http://www.bedfordstmartins.com/hacker/resdoc>, includes the complete text of the booklet and links to all the Internet sites it lists.

Harnack A, Kleppinger E. 2000. Online! A reference guide to using Internet sources. Boston: Bedford/St. Martin's. 272 p.

In addition to providing models for citing and documenting Internet sources, this guide offers extensive advice on navigating the Internet, conducting Web searches, and evaluating sources. The companion Web site, located at <http://www.bedfordstmartins.com/online>, provides models for citing and documenting electronic sources, as well as links to other Web sites for online citation styles.

Literature Cited

Bem SL. 1981. Gender schema theory: a cognitive account of sex typing. Psychol Rev 88:354–64.

Burger J. 1974. Breeding adaptations of Franklin's gull (*Larus pipixcan*) to a marsh habitat. Anim Behav 22:521–67.

Dawkins R. 1976. The selfish gene. New York: Oxford Univ Pr. 368 p.

Eiseley L. 1958. Darwin's century: evolution and the men who discovered it. New York: Doubleday. 449 p.

Ginsberg HJ, Braud WG, Taylor RD. 1974. Inhibition of distress vocalizations in the open field as a function of heightened fear or arousal in domestic fowl. Anim Behav 22:745–9.

Hepler PK, Wayne RO. 1985. Calcium and plant development. Ann Rev Plant Physiol 36:397–439.

Kohmoto K, Kahn ID, Renbutsu Y, Taniguchi T, Nishimura S. 1976. Multiple host-specific toxins of *Alternaria mali* and their effect on the permeability of host cells. Physiol Plant Pathol 8:141–53.

McMillan V, Smith RJF. 1974. Agonistic and reproductive behaviour of the fathead minnow (*Pinephales promelas* Rafinesque). Z Tierpsychol 34:25–58.

Morris D. 1967. The naked ape. Toronto, Canada: Bantam. 366 p.

Nagel M, Oshima N, Fujii R. 1986. A comparative study of melanin-concentrating hormone (MCH) action on teleost melanophores. Biol Bull 171:360–70.

Young CM, Greenwood PG, Powell CJ. 1986. The ecological role of defensive secretions in the intertidal pulmonate (*Onchidella borealis*). Biol Bull 171:391–404.

Index

Symbols and Abbreviations Used in Biology

TERM/UNIT OF MEASUREMENT	SYMBOL/ ABBREVIATION
ångström	Å
approximately	ca. *or* c. *or* ≈
calorie	cal
centimeter	cm
cubic centimeter	cm^3
cubic meter	m^3
cubic millimeter	mm^3
day	d
degree Celsius	°C
degree Fahrenheit	°F
degrees of freedom	df, ν
diameter	diam
east	E
et alia (Latin: and others)	et al.
et cetera (Latin: and so forth, and so on)	etc.
exempli gratia (Latin: for example)	e.g.
female	♀
figure, figures	Fig., Figs.
foot-candle	fc *or* ft-c
gram	g
greater than	>
hectare	ha
height	ht
hour	h
id est (Latin: that is)	i.e.
joule	J
kelvin	K
kilocalorie	Kcal
kilogram	kg
kilometer	km
latitude	lat.
less than	<
liter	l *or* L
logarithm (base 10)	log
logarithm (base *e*)	ln
longitude	long.
male	♂
maximum	max
mean (sample)	$\bar{x}$, $\bar{X}$, $\bar{y}$, *or* $\bar{Y}$

Symbols and Abbreviations Used in Biology

TERM/UNIT OF MEASUREMENT	SYMBOL/ ABBREVIATION
meter, metre	m
microgram	μg
microliter	μL
micrometer	μm
milliliter	mL
millimeter	mm
minimum	min
minute (time)	min
molar (concentration)	mol L^{-1}
mole	mol
month	mo
nanometer	nm
north	N
not (statistically) significant	NS *or* ns
number	nr
number (sample size)	*n*
parts per million	ppm
percent	%
plus or minus	±
probability	*P*
second (time)	s
south	S
species (singular)	sp.
species (plural)	spp.
square centimeter	cm^2
square meter	m^2
square millimeter	mm^2
standard deviation	*s or* SD
standard error	$s_{\overline{x}}$ *or* SE
standard temperature and pressure	STP
versus	vs.
volume	vol
volt	V
watt	W
week	wk
weight	wt
west	W
year	y

Praise for Writing Papers in the B

"This book should be required reading for all students in the biological

Agricultural and Mechanical University

hat it manages to convey, in a wonderfully
nce and information that is indispensable
dvice is complete, clear, and illustrated
ncluding examples of students' writing
redibility to the material."
llege

al Sciences is attractive and easy to read.
s can easily use it as a reference tool. It
view and detail."
ara University

nsive —*Writing Papers in the Biological*
f common sense; to use a term from the

Winthrop University

tion of this book are the clear, easy-to-
, which make it ideal as a regular tool
hman writing projects and as a reference
quire writing or public speaking."

—Laura Mudd, *Barry University*

ISBN 0-312-25857-7

90000

9 780312 258573

BEDFORD/ST. MARTIN'S
www.bedfordstmartins.com